Optimizing Generative AI Workloads for Sustainability

Balancing Performance and Environmental Impact in Generative AI

Ishneet Kaur Dua
Parth Girish Patel

Apress®

Optimizing Generative AI Workloads for Sustainability: Balancing Performance and Environmental Impact in Generative AI

Ishneet Kaur Dua
Dublin, CA, USA

Parth Girish Patel
Dublin, CA, USA

ISBN-13 (pbk): 979-8-8688-0916-3
https://doi.org/10.1007/979-8-8688-0917-0

ISBN-13 (electronic): 979-8-8688-0917-0

Managing Director, Apress Media LLC: Welmoed Spahr
Acquisitions Editor: Celestin Suresh John
Development Editor: Laura Berendson
Coordinating Editor: Kripa Joseph
Copyeditor: Kim Burton

Cover designed by eStudioCalamar

Cover image by Freepik (www.freepik.com)

Distributed to the book trade worldwide by Apress Media, LLC, 1 New York Plaza, New York, NY 10004, U.S.A. Phone 1-800-SPRINGER, fax (201) 348-4505, e-mail orders-ny@springer-sbm.com, or visit www.springeronline.com. Apress Media, LLC is a California LLC and the sole member (owner) is Springer Science + Business Media Finance Inc (SSBM Finance Inc). SSBM Finance Inc is a **Delaware** corporation.

For information on translations, please e-mail booktranslations@springernature.com; for reprint, paperback, or audio rights, please e-mail bookpermissions@springernature.com.

Apress titles may be purchased in bulk for academic, corporate, or promotional use. eBook versions and licenses are also available for most titles. For more information, reference our Print and eBook Bulk Sales web page at http://www.apress.com/bulk-sales.

Any source code or other supplementary material referenced by the author in this book is available to readers on GitHub (https://github.com/Apress). For more detailed information, please visit https://www.apress.com/gp/services/source-code.

If disposing of this product, please recycle the paper

Table of Contents

About the Authors

 **Ishneet Kaur Dua** is an experienced solutions architect specializing in generative artificial intelligence, machine learning, environmental sustainability, and cloud computing. With over eight years of hands-on experience, she excels in designing cost-effective, resilient systems on leading cloud platforms such as Amazon Web Services (AWS), Google Cloud Platform (GCP), and Microsoft Azure. Ishneet started her career at CDK Global, where she worked as a DevOps and Kubernetes engineer, focused on building highly available Kubernetes environments on AWS cloud and on-prem. Passionate about leveraging artificial intelligence (AI) and machine learning (ML) for innovation, Ishneet has expertise in diverse areas, including low-code and no-code ML, computer vision, fine tuning and model customization, natural language processing , recommendation engines, and predictive analytics. She advocates for ethical AI practices, ensuring fairness and transparency in AI systems while making them accessible through open-source initiatives.

As a thought leader, Ishneet shares her insights at global tech conferences, focusing on AI/ML, cloud architecture, and sustainability. She actively mentors women in tech, aiming to inspire and empower the next generation of STEM professionals. Driven by a vision of harnessing technology for positive change, Ishneet is dedicated to building a future where AI creates opportunities for all and addresses complex real-world challenges.

 **Parth Girish Patel** is a seasoned architect with a wealth of experience spanning over 18 years, encompassing management consulting and cloud computing. At Amazon Web Services (AWS), he specializes in artificial intelligence/machine learning, generative ai, sustainability, application modernization, and cloud-native patterns to deliver resilient, high-performance solutions optimized for cost and operational efficiency. Starting his career as a software engineer, Parth transitioned into consulting at Deloitte, where he provided strategic guidance to Fortune companies on their cloud implementation and led intricate enterprise transformations. This diverse background equipped him with a unique blend of business acumen and technical expertise, enabling him to navigate complex digital transformations effectively. As an AWS solutions architect, Parth plays a pivotal role in guiding customers through their cloud journey and AI adoption, offering insights into scalable architectures and implementing end-to-end machine learning solutions. With specialization across leading cloud providers like AWS, Azure, and GCP, as well as proficiency in machine learning skills like natural language processing, computer vision, and predictive analytics, Parth is well-equipped to tackle diverse technical challenges.

Passionate about sustainable AI, Parth advocates for the responsible and ethical use of AI, emphasizing transparency and environmental consciousness. He leverages his leadership skills to mentor teams and individuals, fostering a collaborative and innovative environment aimed at driving a positive impact across organizations and society as a whole.

About the Technical Reviewer

Anandaganesh Balakrishnan is a data engineering and data analytics leader who has held senior leadership roles across the fintech, biotech, and utility domains. His expertise is architecting scalable, reliable, and performant data platforms for advanced data analytics, quantitative research, and machine learning. His current research is AI on unstructured data, large language models (LLMs), generative AI (GenAI), self-service data analytics, and data catalogs.

CHAPTER 1

An Introduction to Generative AI

Generative artificial intelligence, also known as generative AI and GenAI, refers to a category of AI systems that use machine learning models to create new, original content rather than simply analyze existing datasets. Unlike most AI, which focuses on pattern recognition in data to optimize decision-making or predictions, generative AI can synthesize completely new data that has never been seen before by the system. At a basic level, generative AI systems work by learning the underlying patterns and structures within a certain domain of data—whether it be text, images, audio, video, or something else. By analyzing thousands or millions of examples, the AI develops an implicit understanding of what plausible new outputs in that domain could be. It learns features and rules inherent within the data that allow it to emulate and build off what it has seen, going beyond those finite examples. Figure 1-1 demonstrates the hierarchy of artificial intelligence fields.

© Ishneet Kaur Dua and Parth Girish Patel 2024
I. K. Dua and P. G. Patel, *Optimizing Generative AI Workloads for Sustainability*,
https://doi.org/10.1007/979-8-8688-0917-0_1

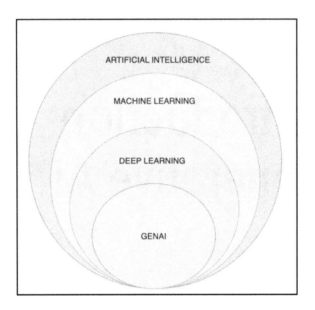

Figure 1-1. *The hierarchy of artificial intelligence fields*

Generative AI and traditional artificial intelligence and machine learning (AI/ML) have fundamentally different goals and approaches. While traditional AI/ML focuses on analysis, classification, and prediction based on existing data, generative AI aims to create and synthesize new artifacts and content, such as text, images, audio, or other media. Generative models like generative adversarial networks and variational autoencoders use techniques focused on learning latent representations and deep neural patterns that capture the essence of data distributions. This allows them to produce novel outputs going beyond their training data. In contrast, traditional techniques like linear regression and convolutional neural networks aim to accurately classify and label examples based on supplied training data.

These differing aims lead to major differences in applications as well. Generative AI enables automated content generation, conversational bots, drug design, and more based on subjective machine creativity. Meanwhile traditional AI improves existing processes via analytics, predictions, and recommendations. Generative AI also poses challenges in accurately evaluating the quality and coherence of machine-created content. So, while generative models greatly expand the scope of artificial intelligence into creative applications, traditional analytical AI retains advantages in explainability and quantifiable performance.

History

Generative artificial intelligence represents an evolution in deep learning techniques aimed at creatively synthesizing novel content. The foundations for generative algorithms emerged in 2014 with the introduction of generative adversarial networks (GANs). Proposed by Ian Goodfellow, GANs comprise two neural networks—generators and discriminators—competing against each other in a zero-sum game framework. By attempting to fool its discriminative counterpart in an adversarial feedback loop, generators continuously improve at constructing realistic synthetic outputs like images or text. Various innovations around conditional, controllable, and progressive GANs followed, achieving remarkable results in automated generation tasks.

Concurrently, other seminal architectures arose, like variational autoencoders (VAEs), diffusion models, and normalizing flows—bringing complementary strengths for generative modeling using probabilistic and iterative techniques. However, a breakthrough arrived in 2018 when Jacob Devlin and his colleagues published the bidirectional encoder representations from transformers (BERT) architecture. Inspired by transformer-based language models, BERT leveraged a vast corpus of textual data through pre-training objectives to learn bidirectional contextual representations. This established a versatile foundation for natural language processing—setting performance records on question answering and other benchmarks while enabling fine-tuning on downstream tasks.

The self-supervised objectives behind BERT soon became a paradigm shift within generative AI. Models like GPT-3 and DALL·E 2 demonstrated transformers' aptitude for zero-shot generalization in text generation and image synthesis solely from pre-training on vast data. Recently, models have rapidly scaled up as computational power grows, passing hundreds of billions of parameters. This massively benefits few-shot learning behavior for creative applications. With ethical considerations around data bias and energy usage, however, the next frontiers lie in knowledge aggregation, causal reasoning, and multimodal generative capabilities through architectures inclusive of varied data modalities. Figure 1-2 shows the model architecture timelines.

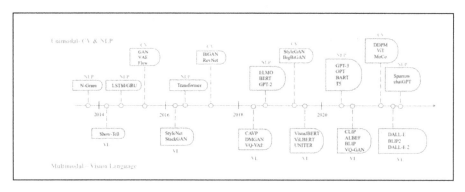

Figure 1-2. *Model architecture timeline*

Types of Generative AI Models

Generative AI models can be categorized as either unimodal or multimodal. Unimodal generative models specialize in one type of media output, such as text, images, or audio. For example, language models like GPT-3 generate written text, while image generation models like DALL-E create original images. Multimodal generative AI combines two or more types of data as both input and output. For instance, the Google Parti (Pathways Autoregressive Text-to-Image) AI model takes an input image and natural language description to generate a matching output image conforming to the text prompt. Other multimodal models may generate images, captions, videos, audio, or integrated media formats. Multimodal generative AI aims to capture connections across modalities, working toward more human-like understanding than unimodal models focused on a single media type. However, unimodal models currently tend to demonstrate higher quality and resolution for their specialized domain. Both varieties of generative AI have distinct strengths suited to different creative applications and tasks. As research continues, we may see more integrated multimodal systems come to the forefront in their ability to synergize different senses and formats of data. As seen in Figure 1-3, generative AI models can be categorized into two types: unimodal models and multimodal models. Unimodal models receive instructions from the same modality as the generated content modality, whereas multimodal models accept cross-modal instructions and produce results of different modalities.

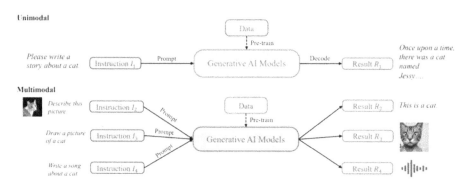

Figure 1-3. *Overview of AIGC*

The Promise and Challenge of GenAI

Generative AI promises a new era of enhanced human creativity, automated efficiency, and on-demand content generation. Recent breakthroughs in deep generative modeling have shown remarkable skill at synthesizing written text, photographic images, video footage, 3D-rendered scenes, musical compositions, and more with increasing levels of realism and coherence. Core algorithmic innovations around adversarial training, diffusion models, and transformer language architectures continue raking in milestones in subjective quality and versatility of machine-made artifacts—whether in simulating logical reasoning or replicating artistic styles.

Businesses already leverage AI generation to augment early ideation phases of product design cycles, accelerate research workflows through scientific literature reviews, and save marketing budgets through automated copywriting and image asset creation. Individual creators fruitfully employ assistance in composing video game visual development assets and drafting interior design 3D renderings matching initial sketches. As barriers to entry lower through user-friendly interfaces, a broader swath of industries stand poised to benefit through increased worker productivity, enhanced customer experiences, and potentially democratized access to once-costly services.

However, formidable challenges around output bias, legal protections, and data transparency must yet be addressed before generative AI can gain societal trust and adoption at scale. Mechanisms for open auditing of datasets and algorithms, watermarking to prevent misuse of machine-made content, plus techniques that empower human guidance of suggestions could help steer technology toward more ethical outcomes. Energy-efficient computing infrastructure is crucial for environmental sustainability as models continue to grow in size. Ultimately, conscientious governance and cross-disciplinary foresight will be instrumental at this critical juncture of inevitable technical improvement.

The Capabilities of Generative AI Systems

Figure 1-4 shows the capabilities of GenAI systems.

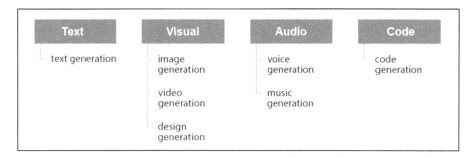

Figure 1-4. *Generative AI tools*

Text Generation

Large Language GenAI models can generate human-like text for a range of applications. After training on millions of web pages and books, they can complete sentences, write paragraphs, summarize texts, translate languages and more based on a few words or sentences of prompts. By learning holistic latent representations, AI models can capture the essence of visual and linguistic concepts from enormous datasets.

Image Generation

Amazon Titan Image, DALL-E, Stable Diffusion, and other image generators can create completely new visual content or digital artworks based on text descriptions. A user can describe a prompt like "an armchair in the shape of an avocado," and the system will generate plausible images matching that description. As generators and discriminators are pitted against each other, GANs grow remarkably proficient at building realistic images from scratch and composing diverse written narratives around specified topics or seeded text.

Creative workflows now leverage these capabilities to ideate book covers matching desired genres, product renderings conforming to initial sketches, icon illustrations fitting branding aesthetics, and magazine layouts populated with AI-generated articles. Conditional GANs allow finer user control by stipulating high-level artistic directions. Meanwhile, diffusion models can rectify image artifacts and enhance text coherence through iterative refinement. As algorithms and data continue to improve, GANs may one day replicate any visual style or literary voice with pinpoint accuracy.

By outsourcing repetitive drudgery, this technology lets humans focus creativity on big-picture goals and final polish. Automating early-stage concepts also reduced costs and environmental waste. At scale, AI generation could even democratize access to services once out of reach for low-budget groups. Still, ethical considerations around data sourcing, artistic credit, and legal usage remain vital as applications broaden. Overall, AI synthesis marks a new frontier in augmented creativity and productivity.

Audio Generation

Projects like Stable Audio by StabilityAI, Google's MusicLM and Anthropic's Claude can generate completely new music in certain genres and styles or synthesize speech based on input texts. This has implications for media production.

Video Generation

Emerging video AI models can generate plausible footage by learning from datasets of existing videos and films. GANs pave an exciting path for AI-powered automated video generation and synthesis. By learning from vast datasets of video content, GAN architectures can isolate and replicate patterns in how real videos are composed—modeling background scenes, objects, textures, motions, and high-level semantics. As models capture increasingly complex video statistics, they can realistically synthesize dynamic scene visuals from compact latent representations. GANs can generate short clips of plausible imagery by "hallucinating" details conditioned on sparse keyframes. As research advances, they may create minutes-long HD videos consistent in style, narrative flow, and crisp detail—whether a nature landscape or a celebrity interview. Beyond modeling observed footage, conditional and controllable GANs could also compose and edit novel video mashups according to customizable specifications. This permits applications like creating synthetic data for augmented reality interfaces, animating 3D rendered storyboards for the media and entertainment industry, cinema pre-visualization, dynamically generating video game environments, or mimicking a client's brand style for advertising and marketing projections. If datasets reflect ethical diversity, GAN-generated video could greatly save production costs and time for multimedia projects while minimizing carbon footprints. As this emergent technology keeps improving, AI-assisted video fabrication appears poised to unlock new creative horizons across industries. This remains an active area of research. StabilityAI has a popular Stable Video Diffusion model that can generate 4sec short video clips from a text prompt.

Programming and AI Code Companions

Generative AI models show immense promise for supercharging code autocompletion and productivity enhancement features in integrated development environments (IDEs). By analyzing extensive source code repositories, deep learning architectures like Codex can capture common syntactic, semantic, and stylistic patterns in how experienced developers write code. This allows IDEs to offer context-aware, diversified, and personalized code snippet suggestions as developers type, enabling them to choose optimal completions that save significant manual typing. As models train on more code with greater complexity, the vocabulary and versatility of these AI-powered autocompletion systems continue to expand. Developers could be assisted with a wider range of languages, libraries, use cases, and coding styles tailored to their personal workflows. Beyond basic completion, generative techniques could also continuously analyze code context as it's typed to provide live recommendations for error fixing, optimization, and other improvements—reducing debugging overhead. Tight IDE integration and customization based on an individual developer's style and project goals are key to realizing these productivity gains. As algorithms and training data improve, AI assistance could become an indispensable asset in accelerating programmer workflows, reducing cognitive load, and enabling faster ideation on the way to more resilient software applications.

Customer Support/Contact Centers

Generative AI is fundamentally transforming contact center capabilities by automating tedious tasks and enabling more efficient, personalized customer support. Powerful language models like GPT-3 and natural processing frameworks can now parse complex customer inquiries, generate conversational responses, and handle common requests instantly. This alleviates contact volume for human agents and resolves simple issues faster without customers languishing on hold.

As algorithms train on more diverse dialogue data tagged by intent, AI assistants grow more adept at comprehending ambiguous input, asking clarifying questions, and directing customers to the optimal solutions or support avenues. Integrating these systems with Customer Relationship Management Tools (CRMs), knowledge bases, and other backend tools allows for precise, consistent responses tailored to each customer's situation. Contact centers could even employ generative models to automatically generate FAQs and microcopy that adapt linguistically across channels. AI promises more satisfying customer experiences overall—especially when blending automated convenience with the emotional intelligence of human agents for sensitive cases. This hybrid approach recognizes not all inquiries can or should be handled by machines alone. Still, intelligent automation of high-volume trivial tickets, web form processing, and appointment bookings gives agents more capacity to resolve pressing customer problems. As long as transparency and ethics guide these AI implementations, generative technology can profoundly augment contact centers in the name of quality, efficiency, and relationship-building.

Enterprises are leveraging large language models (LLMs) across internal and external applications to drive efficiencies and cost savings. Internally, LLMs are being used for automating customer support through conversational bots that can understand the context and respond accurately to questions. They are also helping employees search through vast document troves to find answers faster. The return on investment is calculated from reductions in ticket handling costs and productivity gains. Figure 1-5 shows how enterprises can use LLMs in different use cases.

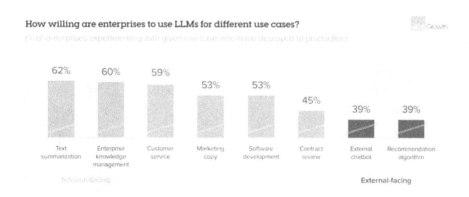

Figure 1-5. *Use cases experiment by enterprises source: a16z survey of 70 enterprise AI decision maker*

Externally, marketing teams are utilizing LLMs to quickly generate high-quality content like social posts and website copy. By partially automating content creation, they are improving velocity and conversion rates. Enterprises' budgets for LLM workloads are based on the potential savings and revenue growth opportunities. More ROI-driven use cases receive higher budgets to scale implementation. The CFO analyzes metrics like interaction rates, conversions, and productivity gains to quantify actual value delivered vs. budgeted amounts. These benchmarks help them decide on continued funding and expansion.

Generative AI by Industry
Media and Entertainment

Generative AI has emerged as a promising innovation for the media and entertainment industry with diverse potential applications that can benefit content creation, distribution, and consumption. See Figure 1-6.

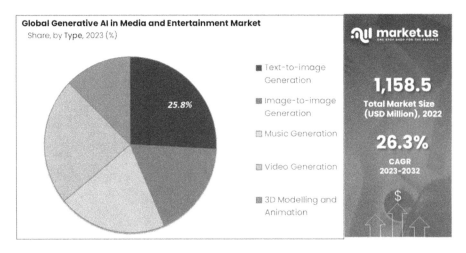

Figure 1-6. *Generative AI use case bifurcation in media and entertainment* https://www.linkedin.com/pulse/predictive-ai-media-entertainment-market-hit-usd-47-billion-g9y8c/

A key benefit is significantly enhanced productivity and reduced costs for content generation. AI algorithms can rapidly generate written scripts, music compositions, visual artwork and more that humans would spend far longer on manually. For instance, tools like Anthropic's Claude can automatically output entire essays, screenplays, or video concepts based on a short prompt. This allows faster ideation and dramatically expands what independent creators or small teams can achieve.

Personalization and recommendation of media experiences is another key application improving customer engagement. Algorithms trained on past preferences can custom-tailor and dynamically adapt entertainment offerings to the precise interests of every individual. Startups like Humai are bringing niche, tailored recommendations to podcasts, while AI could alow infinite personalized playlists in music services.

Generative AI also enables entirely new mediums and creativity-augmenting tools to empower human artists. Systems like Stability AI's Stable Diffusion & Google's Imagen produce striking images from text, while Facebook and Sony have revealed lifelike video rendering. Such innovations provide creative launching pads for multimedia projects constrained previously by skill sets and labor bandwidth. Democratizing creation further, TikTok has invested heavily in AI editing assistants any user can leverage for professional-grade short video output.

Procedurally generating endless landscapes, characters, and questlines removes the limitations of pre-scripted experiences. Overall, generative AI introduces step-change productivity, personalization, and creative enhancements across media verticals from writing to music, virtual worlds, and more. It distills engaging aspects of existing content into formulas while augmenting human expression in pioneering directions together, promising a more vibrant connection between creators and audiences globally. Regulatory challenges around copyright and mature content remain; however, if addressed responsibly, the paradigm shift underway may profoundly uplift industries and societies alike.

Gaming

Generative AI stands to profoundly expand gaming experiences through smarter content creation, more immersive worlds, and data-driven engagement optimization—promising more vibrant, responsive player connections industry-wide if adoption balances innovation with responsibility. See Figure 1-7.

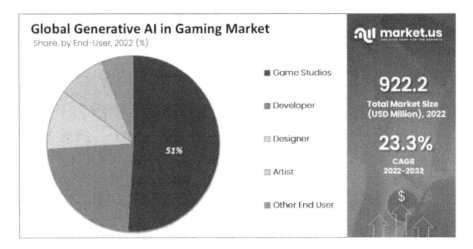

Figure 1-7. *Generative AI gaming market* $https://www.linkedin.$
$com/pulse/future-cloud-gaming-market-trends-growth-$
$projections-markets-us-rwwgc/$

A key opportunity is significantly enhanced game development productivity by automating rote tasks. Generative techniques can rapidly generate game assets like character models, textures, vegetation, and level layouts to accelerate worldbuilding based on high-level creative direction. For instance, tools like GameBake autogenerate modular 3D environmental components customizable into vast unique landscapes compared to intensive manual effort previously.

Such generative content also enables dynamic in-game experiences reacting uniquely to each player. Concepts like infinite quest generation via AI adaptively craft personalized narrative arcs, enemies, and rewards by modeling individual behavior. Startup Anthropic trains AI models as in-game characters able to hold free-flowing conversations reachable no other way. This level of personalization and expression promises next-generation immersion.

Under the hood, generative algorithms optimize metrics like long-term engagement through data-driven experimentation impossible manually. Features including dynamically adaptive difficulty tuning via DeepMind and retention improvement by startups like GameOn demonstrate techniques benefiting developers and players mutually via optimized challenges.

Responsible questions remain around content oversight, IP protection, and potential market consolidation if dominating platforms emerge. Yet balanced with ethical safeguards, creatively democratizing tools underpinning more responsive, personalized, and data-optimized interactive experiences stand to provide quantum leaps for gaming and adjacent metaverse applications through sustainably expanded technological capabilities. The results may profoundly deepen creative connections at a global scale when the human priorities of security, expression, and cooperation lead in parallel with accelerating innovation.

Marketing and Advertising

Generative AI holds significant potential to transform marketing and advertising with a diverse range of impactful applications—from ideation to personalization, creative generation, and execution. These emerging innovations promise enhanced consumer engagement, rapid productivity gains, and novel advertising formats to connect brands with increasingly diverse audiences in an evolving digital landscape.

One of the most promising applications is using generative techniques to develop creative marketing assets like display ads, social posts, or commercial storyboards. Algorithms trained on brand guidelines and past successful creative material can rapidly output entirely new options for campaigns to consider, significantly boosting ideation velocity compared to sole reliance on human teams.

Generative AI also allows template-driven dynamic creative optimization, scaling mass personalization of messaging. Brands can design ad templates with parameters for features like product imagery that algorithmically vary in real-time based on individual viewer data and response, while adhering to brand safety. Such tailored ads deliver two to three times higher engagement, demonstrating the technique's efficiency for advertisers and meaningfully customized consumer experiences.

Powerful emerging techniques also show potential for AI to execute elements of campaign activation via orchestration of channels, budget allocation across segments, and bidding into ad auctions optimized for KPIs.

All considered, generative AI introduces overdue enhancements to marketing operations from ideation to execution via superior personalization, creative automation, and beyond. Though policy and job impact concerns remain with any technological shift, responsibly leveraged, Generative techniques may allow brands to communicate more resonantly across exponentially diversifying channels and consumer groups through intelligently amplified capabilities. The result can be sustainably accelerated progress resonating positively across commercial ecosystems and touchpoints when priorities appropriately balance value generation alongside responsible disruption.

Automotive

One major application is significantly faster design iteration by generating CAD models, prototypes and photorealistic renderings to evaluate styling options in a fraction of current timelines before physical modeling. For instance, AI startup Genesis has demonstrated producing exterior and interior renderings for review in under 30 seconds. Such agility allows more concepts to be considered and selected using actual data. See Figure 1-8.

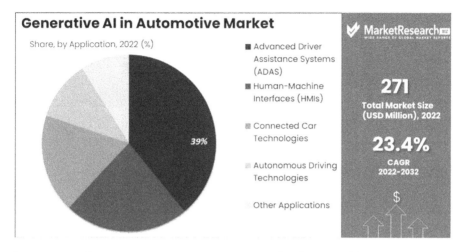

Figure 1-8. *Generative AI automotive market* `https://`
`marketresearch.biz/report/generative-ai-in-automotive-market/`

Generative techniques likewise show enormous potential in enhancing production via AI-generated blueprints optimized for manufacturability. Algorithms can rapidly iterate mechanical, electrical, and software configurations for the highest quality and efficiency, mapping interdependencies impossible to visualize manually. Initial implementations have already reduced machine downtime, scrap, and overhead through simulation.

Finally, generative deep neural networks are poised to transform autonomous system robustness and safety. Approaches like cascaded networks intelligently weighing multiple driving policy outputs or generative adversarial simulation can minimize risk profiles beyond human capability through billions of virtual test miles. Though the adoption pace remains uncertain, hands-off wheel maturity could accelerate given already immense flexibility gains in design and production from AI leverage.

In total, implemented ethically, generative technologies promise step-change gains in automotive development velocity, manufacturing optimization, and vehicle autonomy, promising consumer and enterprise value chain uplift when priorities ensure responsible innovation curbs excesses through employee and community empowerment. But practiced narrowly, risks remain of excessive disruption without adequate transition support. If embraced cooperatively however, a sustainable renaissance helping mobility achieve its highest purpose beckons within reach.

Manufacturing and Industrial

Generative AI heralds abundant opportunities to transform manufacturing through enhanced design, operational efficiencies, and workforce augmentation that—properly directed through ethical governance—promise vastly expanded sustainable progress and economic access across the industry. See Figure 1-9.

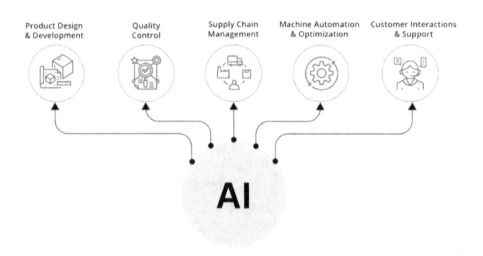

Figure 1-9. *Generative AI in manufacturing* https://medium.com/ @NeuralPit/discover-the-future-of-manufacturing-how-generative-ai-is-redefining-efficiency-and-innovation-239d6a7359ae/

Design automation constitutes a primary application area via intelligent CAD tools. By algorithmically exploring intricate combinations of shapes, materials, and subcomponents manually, generative techniques like topology optimization stand to rapidly enhance prototyping, lightweighting efforts, and performance modeling. Toyota and Airbus already utilize such methods reducing parts and weight while strengthening integrity.

Generative AI also introduces step-change gains for process quality assurance and control on production lines. Computer vision techniques can continuously validate manufacturing steps non-invasively in real-time to minimize errors or drift overruns, while natural language interfaces democratize interaction with equipment across skill levels. Initial pilots demonstrate sizable scrap and rework reductions achievable augmenting human insight.

Perhaps most critically, generative approaches promise to unlock integrated supply chain transparency, predictive risk management, and responsibility across intricate multinational value webs. Simulating cascading implications of disruptions while optimizing continuity, ethical sourcing, equity, and environmental considerations systemwide remains essential for stable access to goods globally and can be exponentially enhanced via AI leverage.

In total, when governance and workforce participation responsibly guide technological integration, generative methods offer a multiplying force for human ingenuity, realizing more abundant, accessible, and sustainably manufactured products worldwide—raising standards of living while safeguarding futures for all through compassionately accelerated innovation scaled for cooperation not merely disruption.

Semiconductors

One major application area constitutes Integrated Circuit (IC) layout generation and verification. Companies like Synopsys seek to train generative adversarial networks that intelligently explore immense design spaces to output optimized chip configurations matching specified parameters beyond human engineers. Such intelligent automation can shrink development cycles for advanced nodes from years to months while minimizing power and maximizing yield. See Figure 1-10.

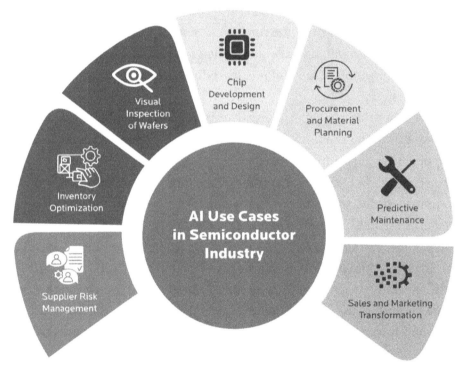

Figure 1-10. *Generative AI in the semiconductor industry* https://www.birlasoft.com/articles/ai-semiconductor-industry-use-cases-and-benefits

Fabrication process tuning through simulation also stands to benefit enormously from data-driven automation. Algorithms can rapidly identify equipment adjustments for target wafer quality by modeling complex chemical interactions during deposition or etching that influence interdependent variables. Initial implementations already report 30%–40% faster ramps and defect reductions.

Finally, supply chain continuity represents a crucial opportunity to apply generative risk prediction and mitigation methods. Simulating potential bottlenecks and guiding intelligent capacity distribution, inventory buffers and purchasing contingency worldwide would prove enormously valuable by sustaining output despite more frequent disruptions ahead. Such continuity holds special relevance for economies depending on electronics access.

Generative AI could profoundly transform the semiconductor landscape toward more agile, resilient, and sustainable electronics innovation ecosystems when governance ethically distributes benefits equitably among workforce and community stakeholders. If technology integration instead narrowly concentrates value, risks will multiply. But directed toward cooperative human progress, generators promise a democratization of capability unlike any preceding era—one optimized responsibly through compassionate governance and priority.

Retail and Ecommerce

Generative AI introduces the enormous potential to enhance retail and ecommerce across opportunity areas from demand forecasting to personalized recommendations and intelligent supply chains—collectively promising more relevant, reliable commerce through properly governed technology integration. See Figure 1-11.

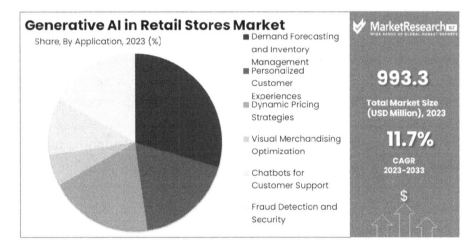

Figure 1-11. *Generative AI in the retail stores market*
`https://marketresearch.biz/report/generative-ai-in-`
`advertising-market/`

One major application constitutes leveraging generative algorithms for significantly more accurate demand modeling and optimization. By synthesizing immense datasets spanning economic indicators, search trends, purchasing behavior and real-time events, AI can continuously output updated sales projections for inventory and operations planning exceeding human capability. Early retail implementations already report 40%–60% forecast accuracy improvements at scale.

Generative customer engagement via personalized recommendations and messaging also promises tremendous uplift by tailoring interactions to individual preferences. Algorithms trained on browsing history, purchase data, and demographics can dynamically compile customized product bundles, pricing, and creatives for every consumer touchpoint, from ads to sites and stores. Companies like Bloomreach see higher conversion gains possible through bespoke content and offers.

Behind the scenes, generative supply chain optimization will prove crucial for retail continuity by preempting disruptions through simulation.

Scenarios modeling component availability, manufacturing conditions, and logistics flows globally can prescribe mitigation actions before bottlenecks emerge across intricate, interdependent value chains. What-if analysis helps strategically inform expansion, partnerships, and more to sustain access.

In total, applied conscientiously, generative innovations may profoundly expand retail's capability to responsively satisfy consumer needs worldwide through enhanced intelligence across planning, engagement, and network resilience. But without adequate data and job transition safeguards, such automation risks disruption. If embraced cooperatively, however, as complements to human ingenuity, generators can exponentially uplift equitable value creation.

Financial Services/Fintech

Generative AI promises far-reaching transformations across fintech—from democratizing access through automated advice to revolutionizing risk modeling and combating fraud with intelligence exceeding human limitations. Responsibly applied, generators can uplift financial inclusion, efficiency, and continuity, though they require balanced governance to ensure equitable innovation. See Figure 1-12.

How Generative AI Can Add Value to Financial Services' Tasks

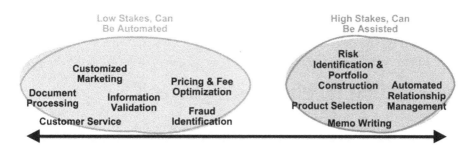

Figure 1-12. *Value-add of generative AI in financial services*
https://fintechtakes.com/articles/2023-02-17/generative-ai/

A crucial opportunity exists around using generative techniques to provide customized financial guidance and product recommendations for underserved groups previously lacking affordable access. Models trained on sparse transaction histories can still output tailored savings, investment, and credit options accounting for unique needs and risk tolerance even with limited history—helping bring fair services to more.

Fraud prevention and detection similarly stands to make quantum leaps in both consumer and institutional fintech by applying generators' pattern intuitions exceeding people. Identifying novel fraudulent accounts or transaction indicators across datasets promises far faster adaptation speed against the next unpredictable threat vector before losses multiply. High-frequency decision automation adds an advantage.

Core credit and risk assessment engines stand transformation by algorithms correlating immense alternative datasets like supply chains, transportation systems, and e-commerce to anticipate areas of emerging risk or opportunity in real time, which is unattainable manually. Financial models growing more representative may better serve communities should governance guide innovation equitably.

Applied conscientiously generative AI could profoundly expand access, security, capability, and value across global fintech, radically improving lives, but only if participant rights led to technological integration. Without adequate safeguards and support for those disrupted, underlying human progress risks subordinate to efficiency gains but directed responsibly, generators offer exponentially uplifting tools for the disadvantaged worldwide once excluded.

Information Technology/High-Tech

Generative AI promises abundant transformative IT solutions spanning enhanced software engineering, smarter infrastructure management, and better cybersecurity protections that offer radically accelerated digital capabilities—if thoughtful governance and workforce policies responsibly guide adoption toward equitable innovation.

A major opportunity exists in applying generators to automate routine yet intricate coding tasks. Algorithms trained on huge codebases can rapidly output entirely new functions to specification, saving developers months of work better spent creatively problem-solving. Startups like Anthropic claim 10x gains in coder productivity today possible while democratizing participation. Responsible data usage and rights protections remain vital, however to ensure representative security in implementation.

IT operations stand transformation too from AIOps scaling predictive infrastructure orchestration. By manually synthesizing telemetry across servers, networks, and apps, generative algorithms can preemptively optimize configurations, resource allocation, and failover procedures for resilience and efficiency, exceeding current practices reliant on reactive responses. Early proof of concepts shows more than 40% cost and incident reductions achievable.

Finally, enhanced cybersecurity constitutes an enormous opportunity where generating advanced threat detection models and automating response procedures promises to combat risks far faster than adversaries can evolve attacks. Techniques synthesizing patterns across entire firms stand to identify novel intrusions early and prescribe tailored containment measures benefiting all while advancing understanding.

In total, applied conscientiously, generative methods can profoundly uplift IT capabilities, efficiencies, and access at a global scale—but require balanced governance emphasizing workforce welfare and cooperative innovation. Hence, all communities benefit from expanding technological abundance rather than excessive disruption. If embraced responsibly, generators offer tools to unlock the next era of ethical digital elevation.

How Enterprises are Buying and Building GenAI

In 2023, generative AI saw explosive growth in the consumer market, rapidly reaching over $1 billion in consumer spending. In 2024, the enterprise market presents an even bigger revenue opportunity, potentially many times larger.

Last year, while consumers embraced AI chatbots and content creation tools, enterprise adoption seemed limited to a few obvious use cases and "wrapping" GPT models into new products. Some doubted whether generative AI could truly scale for business uses. However, after speaking with dozens of enterprise leaders, it's clear that attitudes and investments have shifted dramatically in just the last six months.

Although they still have some concerns, these leaders are nearly tripling their AI budgets, expanding use cases for smaller open-source models, and transitioning more pilot projects into production. This represents a massive opportunity for AI startups that (1) build solutions tailored to strategic enterprise AI initiatives while addressing pain points and (2) shift from services to scalable products.

To help founders understand this enterprise opportunity, we've outlined 16 key considerations around resourcing, models, and use cases that reflect how enterprise leaders think about and adopt generative AI. Building any enterprise product requires a deep understanding of customer priorities and roadmaps. These insights aim to help AI startups build what enterprise customers need in this new wave of investment.

Environmental Sustainability in IT Workloads

Environmental sustainability refers to the responsible interaction with the environment to avoid depletion or degradation of natural resources and allow for long-term ecological quality. It involves maintaining the factors and practices that contribute to the quality of the environment on an ongoing basis. The United Nation's Sustainable Development Goals include multiple key tenets, as shown in Figure 1-13.

Figure 1-13. *The UN's Sustainable Development Goals*

The following describes the key principles of environmental sustainability.

- Natural resource conservation: Using renewable, reusable, and recyclable resources efficiently to preserve supplies for future generations and reduce pollution. Examples include conserving water, minimizing fossil fuel use, and sustainable forestry.

- Biodiversity protection: Protecting the variety of plant and animal life through conservation efforts. Maintaining biodiversity allows ecosystems to adapt and be resilient.

- Waste reduction: Reducing, reusing, and recycling waste to minimize landfill contributions by following the "reduce, reuse, recycle" hierarchy.

- Climate change mitigation: Implementing practices that reduce the emissions of heat-trapping greenhouse gasses and lower an individual, business, or country's carbon footprint through energy efficiency, clean energy adoption, and forest protection.

- Sustainable development: Meeting current economic, social, and environmental needs without limiting the ability of future generations to meet their own needs. Making sustainability a part of business and community planning decisions.

These key pillars require long-term systems thinking, global collaboration across sectors, localized grassroots action, and technological innovation to create environmentally sustainable practices for current and future generations.

The major cloud service providers, including Microsoft Azure, Google Cloud Platform, Amazon Web Services, and others, have a variety of sustainability initiatives aimed at reducing environmental impact. These providers have pledged to shift their operations to 100% renewable energy usage and achieve carbon neutrality or negativity by certain target years. They are designing and building innovative water-less and energy-efficient data centers powered by features like liquid cooling and AI optimization. Cloud providers also offer customers carbon footprint estimators and sustainability calculators to help track emissions from cloud workloads, as well as sustainability advisory services that assist with migrations and optimizations for reducing resource usage. Through hardware and software efficiencies, renewable energy procurement, and working closely with customers on sustainable best practices, the leading cloud players are helping address sustainability for the rapidly growing IT industry.

Let's look at one of the Cloud Service Providers (CSPs) and what they do to ensure cloud workloads are architected well for sustainability. Amazon Web Services (AWS) addresses environmental sustainability in key ways.

- Renewable energy and carbon neutrality commitments: AWS has made public commitments to reach 100% renewable energy usage for its global infrastructure by 2025 and net-zero carbon by 2040. The company has launched multiple large-scale solar and wind farms to work toward these goals.

- Efficiency of data centers: AWS designs and operates its data centers to be as energy efficient as possible through optimizing cooling systems, server workload configurations, building temperature set points, and more. These efforts help reduce overall electricity usage.

- Sustainable hardware lifecycles: AWS uses sustainable electronics practices around IT hardware lifecycles. This includes reusing decommissioned server equipment wherever possible, responsibly recycling electronic waste, and moving toward more energy-efficient chips within its cloud computing hardware.

- Tools for customers' carbon footprints: AWS provides customers with tools to estimate, monitor, and reduce the carbon footprints of their cloud workloads and resources. Features like the AWS Carbon Footprint tool give analytics around emissions.

- Industry partnerships and standards: AWS partners with sustainability-focused organizations like The Climate Pledge to advance best practices. It also aligns with standards like the ISO 14001 environmental management standard across its operations.

In summary, from its infrastructure operations to the tools it provides customers, AWS aims to lead the cloud industry in environmental sustainability efforts and carbon reduction actions. It has set ambitious goals and innovates across areas from renewable energy to hardware efficiency. Another key initiative that AWS undertook to help customers deploy cloud-native workloads that are both well-architected and environmentally sustainable was launching sustainability as the sixth pillar in its well architected framework. This provides customers with best practices across critical areas like security, reliability, and performance to build robust cloud architectures.

One of the key pillars within this framework focused specifically on sustainability highlights steps for maximizing cloud efficiency, thereby minimizing environmental impact. AWS outlines guidance centered on modeling emissions through dedicated tools, right-sizing infrastructure via autoscaling capabilities and savings plans, regularly deprovisioning unused resources, and keeping architectures current by leveraging AWS' latest energy-optimized hardware and services. By incorporating the sustainability pillar into the planning process, cloud architects can construct systems that achieve economic and operational goals while monitoring and optimizing their carbon footprints through measurable tracking and impact reduction mechanisms available natively through AWS. See Figure 1-14.

| Operational Excellence | Security | Reliability | Performance Efficiency | Cost Optimization | Sustainability |

Figure 1-14. *AWS Well-Architected Framework 6 pillars*

The Environmental Impact of AI Workloads

Artificial intelligence systems and models are increasingly prevalent, powering everything from product recommendations to self-driving vehicles. As AI adoption accelerates across industries, these systems require massive amounts of data and computing resources for development and continuous learning, leading to fast-rising energy demands. Without consideration of sustainability, the environmental impacts of AI in emissions, e-waste, and resource depletion could quickly outweigh the benefits. Establishing sustainable practices is crucial for balancing AI's societal value and ecological consequences.

Training AI models, especially large neural networks used in deep learning, involves substantial computational requirements. State-of-the-art models contain billions of parameters, necessitating clusters of graphics cards or dedicated AI chips to learn complex behaviors from huge sets of data. Generative AI models like DALL-E can emit hundreds of thousands of pounds of carbon dioxide equivalent during development. Collectively across companies, applications, and use cases, unchecked AI growth on current hardware can contribute enormously to climate impacts. Incorporating sustainability research and awareness into every stage of building AI systems establishes positive norms against unsustainable exponential scale. Researchers must explore alternative algorithms and architectures for efficiency from the outset before resource-intensive deployment. To provide a sense of change in scale, model sizes have grown 1,600 times since 2019 and are expected to continue to grow further, as shown in Figure 1-15.

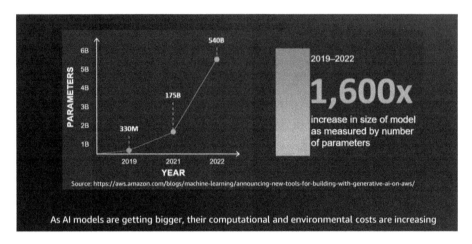

Figure 1-15. *Timeline of an increase in model parameter size*

The notion of "AI waste" also directly harms sustainability through unnecessary emissions and electronics waste from suboptimally designed models. When AI systems are inefficiently constructed, key metrics around performance, fairness, robustness, or compute requirements suffer, leading to models that must be fully redeveloped. Such redundant experimentation and versioning unsustainably burn through irreplaceable hardware components during training. Disposing of partially useful models meets some goals but not others, which creates e-waste. Sustainable development practices emphasize reuse and modularization to avoid duplication and excess waste. As examples, centralized model banks, continual model retraining, and compute-sharing techniques present solutions companies and researchers should embrace. Improving AI's efficiency, accountability, and component reuse is imperative for sustainable progress rather than solely maximizing predictive accuracy metrics through brute-force scale.

Beyond ecological impacts, sustainable AI also entails ensuring technologies developed today do not disproportionately compound existing harms against already marginalized groups in society. As machine learning algorithms and autonomous systems take key roles in high-impact domains like finance, healthcare, and security, they raise complex ethical questions around embedded biases that can violate civil liberties or exacerbate inequality. For instance, predictive policing tools relying on flawed data have faced widespread allegations of racial prejudice. Over-emphasis on profits over diversity has led to alarming demographic skews in many AI datasets and teams shaping societally influential technologies. Achieving ethical AI requires questioning norms around what business efficiencies should compromise. Incorporating voices from marginalized backgrounds directly into development workflows, prioritizing representativeness in data collection, and enabling transparency through external audits present actionable ways toward sustainable AI ethics alongside ecological sustainability.

In conclusion, balancing AI's great promise for progress against its potential unintended harms obligates both bottom-up and top-down transformations around how these exponentially powerful technologies are built. Researchers optimizing algorithms for efficiency and exploring alternative paradigms can establish precedents that scale drives underlying merit rather than solely predictive accuracy or profitability. Policymakers enacting guardrails aligning private interests with public welfare can attenuate negative externalities from unsustainable practices. Companies building AI capabilities wield immense power to self-regulate by intrinsically designing transparency, oversight, and recyclability into their systems. Through shared responsibility across stakeholders and dimensions like engineering, ethics, and ecology, AI can uplift society equitably today and for future generations.

A recent study analyzed the immense energy usage and subsequent carbon footprint of training an extra-large 175-billion-parameter AI model. Powering and running the computational processes to develop this complex neural network consumed a staggering 1,287 megawatt hours of electricity. When accounting for the source of this electricity, the model's training resulted in the emission of 502 metric tons of carbon into the atmosphere. To put that in perspective, the carbon emitted equals that from operating 112 standard gasoline-powered cars for an entire year.

Beyond just the training phase, data centers that host AI systems have huge electricity demands, taking up 2% of total energy utilization in the United States. Some estimates show these specialized facilities require anywhere from 10 to 50 times more energy per square foot than typical offices. Researchers have also compared the energy needed to power certain AI applications to the equivalent of drinking multiple bottles of water. As models scale up in parameters and capabilities, their voracious consumption of computing resources directly translates into more indirect environmental impacts through carbon emissions and resource usage.

On top of emissions, there are further sustainability considerations around electronic waste generated from discarded AI hardware and infrastructure. This waste contains hazardous chemicals like lead, mercury, and cadmium that can pollute soil and water if not properly disposed of. Responsibly managing this complex technology poses environmental challenges now and into the future as AI continues to expand.

Generative artificial intelligence models like DALL-E, GPT-3, Claude, and Stable Diffusion have brought impressive advances in AI capabilities. The following image shows us the CO_2 emissions benchmarks for some common usage scenarios. The CO_2 emissions footprint is high for training an AI model, even when compared to air travel or someone's carbon footprint in their entire lifespan. This is shocking and, hence, why we need to start thinking about the potential damage this technological shift can cause to our planet, making it even more important that we optimize our resource consumption and be mindful of the environmental repercussions of GenAI. See Figure 1-16.

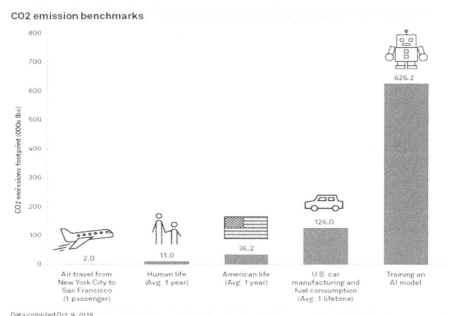

Figure 1-16. *CO2 emission footprint comparison* https://www.forbes.com/sites/glenngow/2020/08/21/environmental-sustainability-and-ai/

Generative AI models utilize techniques that involve pre-training a very large neural network on massive datasets before fine-tuning it for downstream tasks. The computational resources required for pre-training are immense. According to one estimate, training a model like GPT-3 can emit as much as 626,000 pounds of carbon dioxide equivalent—nearly five times the lifetime emissions of the average American car, including fuel usage.

The main environmental costs stem from the hardware, electricity, and data center infrastructure needed to train and run these models. Training occurs on specialized hardware like graphics processing units (GPUs), which consume considerable electricity. Data centers that house clusters of this hardware have huge energy and emissions footprints.

As generative models grow ever larger and more capable, their training requires exponentially more computing resources. If left unchecked, the environmental impacts could quickly balloon. More research is needed to quantify costs and investigate solutions like developing standardized benchmarks, open-sourcing models, and utilizing alternative model architectures. Overall, the AI community must prioritize model efficiency and sustainability alongside performance gains from scale as generative models continue proliferating.

Summary

Generative AI represents a paradigm shift in artificial intelligence focused on creating new, original content rather than simply analyzing existing data. These systems learn the underlying patterns and representations within different data domains like text, images, and audio to synthesize novel artifacts, mimicking and expanding upon their training data. Key innovations driving this revolution include generative adversarial networks, variational autoencoders, and large transformer language models exemplified by GPT-3, Anthropic Claude, StableLM, Amazon Titan and more. Generative AI unlocks new frontiers of subjective machine creativity with far-reaching applications across industries—from automating content generation and personalizing customer experiences to accelerating product design, optimizing manufacturing, and enhancing cybersecurity.

However, as these generative models exponentially increase in size and complexity to become more capable, their immense computational requirements raise sustainability concerns around energy usage and carbon emissions. Training state-of-the-art systems can emit hundreds of thousands of pounds of carbon dioxide equivalent, more than some cars over their lifetime. Improving model efficiency, developing more sustainable architectures, reducing redundant experimentation, and prioritizing reusability present crucial objectives. Establishing ethical practices around representative data, fairness in development, and transparent governance is vital to ensure generative AI uplifts all communities equitably while minimizing its ecological impact. Realizing this powerful technology's full potential hinges on thoughtfully balancing its societal benefits against environmental and societal costs.

The next chapter dives into the cutting-edge world of generative AI, exploring its architectures, uncovering its pitfalls, and mastering techniques for optimized, responsible, and collaborative AI systems.

CHAPTER 2

The Fundamentals of Efficient AI Workload Management

There has been a surge in the demand for efficient and scalable artificial intelligence (AI) workloads. As AI models continue to increase in complexity and size, managing and optimizing the computational resources required for training and deploying these models has emerged as a critical challenge. This chapter aims to elucidate the essential principles and practices that enable organizations to effectively harness the power of AI while maximizing resource utilization and minimizing operational costs.

Efficient AI workload management is crucial for organizations seeking to leverage the transformative potential of AI technologies. It involves a holistic approach that encompasses hardware optimization, software optimization, data management, and resource allocation strategies. By understanding and implementing these fundamentals, organizations can ensure that their AI workloads run smoothly, efficiently, and cost-effectively, regardless of the scale or complexity of the AI models involved.

One of the key aspects of efficient AI workload management is the utilization of specialized hardware accelerators, such as graphics processing units (GPUs) and tensor processing units (TPUs). These hardware

© Ishneet Kaur Dua and Parth Girish Patel 2024
I. K. Dua and P. G. Patel, *Optimizing Generative AI Workloads for Sustainability*,
https://doi.org/10.1007/979-8-8688-0917-0_2

components are designed specifically to handle the computationally intensive tasks associated with AI workloads, enabling faster training and inference times while reducing energy consumption and operational costs.

In addition to hardware optimization, efficient AI workload management also involves optimizing software components, including deep learning frameworks, libraries, and model architectures. Organizations can maximize the performance and efficiency of their AI workloads, ensuring optimal resource utilization and minimizing computational bottlenecks by leveraging techniques such as batching, parallelism, and model compression.

Effective data management is another critical component of efficient AI workload management. This includes implementing efficient data loading and preprocessing pipelines, leveraging data caching and prefetching techniques, and employing data compression and quantization strategies to reduce memory footprint and accelerate data transfer.

By understanding and implementing the fundamentals of efficient AI workload management, organizations can unlock the full potential of AI technologies while ensuring scalability, cost-effectiveness, and operational efficiency. This chapter delves into these fundamental concepts, providing readers with a comprehensive understanding of the principles and practices that underpin efficient AI workload management.

Understanding AI Workload Characteristics

An AI workload refers to the computational tasks and processes involved in developing, training, and deploying an AI or machine learning (ML) model. It encompasses the entire lifecycle of an AI system, from data collection and preprocessing to model training, evaluation, inference (making predictions or generating outputs), and continuous monitoring and optimization of the deployed model in production. Figure 2-1 illustrates the ML lifecycle.

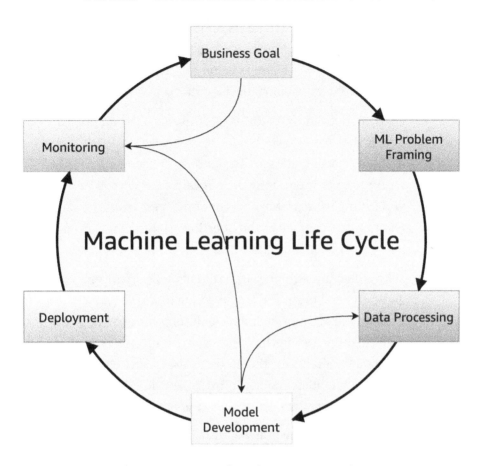

Figure 2-1. *Machine learning lifecycle*

The ML lifecycle is a cyclic, iterative, structured approach that outlines the various stages involved in building and deploying an ML model. It provides a framework for managing the end-to-end process of developing and maintaining AI systems. The typical ML lifecycle, presented in Figure 2-1, consists of the following stages.

- Business goals: The process begins with a business goal. Business goal identification is the most crucial phase of the ML lifecycle. An organization considering

ML must have a clear understanding of the problem to be solved and the business value to be gained. The ability to measure business value against specific business objectives and success criteria is essential.

- Problem definition: In this stage, the specific problem or task that the AI system needs to solve is clearly defined. This includes understanding the business requirements, identifying the relevant data sources, and setting measurable goals or objectives. This is where we ask ourselves the question- Is this a machine learning or GenAI problem?

- Data collection and preparation: This stage involves gathering and preprocessing the data required for training the ML model. It may include tasks such as data cleaning, formatting, and augmentation to ensure the data is suitable for the model. Studies have showed that data scientists spend almost 70% of their time at this stage doing data pre-processing.

- Model development: During this stage, the appropriate ML algorithm or model architecture is selected based on the problem and available data. This may involve exploring different model types (e.g., supervised, unsupervised, or reinforcement learning) and experimenting with various architectures and hyperparameters.

  - Model training: In this stage, the selected ML model is trained on the prepared data. This process involves feeding the data into the model and adjusting its internal parameters (weights) to minimize the error or optimize a specific objective function.

- Model evaluation: After training, the model's performance is evaluated using a separate test dataset or other evaluation metrics. This stage helps assess the model's accuracy, generalization capability, and potential biases.

- Model deployment: If the model meets the desired performance criteria, it is deployed into a production environment. This may involve integrating the model into existing systems, setting up infrastructure for serving predictions or generating outputs, and implementing monitoring and maintenance processes.

- Model monitoring and maintenance: Once deployed, the model's performance and prediction results are continuously monitored, and any issues or performance degradation are addressed. This may involve retraining the model with new data (data augmentation), fine-tuning hyperparameters, or updating the model architecture as needed.

It is important to note that the ML lifecycle is an iterative process, and the stages may be revisited or repeated as necessary to improve the model's performance or adapt to changing requirements.

In the context of AI workloads, understanding the ML lifecycle is crucial because it helps identify the computational resources and infrastructure needed at each stage. For example, the model training stage may require powerful hardware (e.g., GPUs or TPUs) and distributed computing resources, while the inference stage may prioritize low latency and efficient resource utilization.

By breaking down the AI workload into the different stages of the ML lifecycle, it becomes easier to optimize and manage the computational resources effectively, ensuring efficient and sustainable AI development and deployment.

Why the Generative AI Lifecycle Is Different

The generative AI lifecycle exhibits several unique characteristics and challenges that differentiate it from traditional software development or other AI/ML development processes. These distinctions primarily arise due to the nature of generative models, their operational requirements, and use cases. While the lifecycle stages remain similar, generative AI presents a different set of challenges. Figure 2-2 highlights the various phases of the generative AI lifecycle.

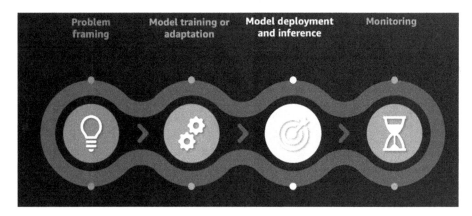

Figure 2-2. *Generative AI lifecycle*

Problem Framing

Clearly define the problem you want to solve and make sure it is a problem that needs generative AI. To better understand the nature of the problem, the following questions can help you decide if generative AI is the correct way to solve it.

- Is the problem related to generating new content, such as text, images, audio, or other data?

- Does the problem require creativity, imagination, or the ability to produce novel and diverse outputs?

- Can the problem be framed as a generative task, where the AI model understands and learns patterns from existing data and generates new answers?

- Determine what success looks like. Define the key performance indicators (KPIs) to measure success.

- Does the problem require domain or industry expertise, or is it trying to solve a generic multifaceted problem?

There are scenarios where you will realize not the entire problem but the specific aspects of the problem that can be addressed with generative AI. Once the problem is defined, selecting the right foundation model is key to minimizing resource usage when building systems using generative AI. First, assess different GenAI models' capabilities and limitations and ensure the solution aligns with the desired outcomes. As the field evolves, domain-specific models, region/localization-specific models, and industry-specific models will emerge. This stage involves researching existing techniques, identifying target outcomes, and establishing evaluation metrics. There are multiple considerations such as model type (image or language), model size, language, domain, fine tune-ability that go into model selection and if done properly, reduces the need to customize the model down the pipeline.

Architecture and Customization

To define the best architecture which suits your problem statements there are few commonly used architectures. Most of problems fall under these categories. Remember these are the very early days of generative AI; new methods will likely emerge, but they will be extensions of these categories. Figure 2-3 features patterns used with generative AI.

45

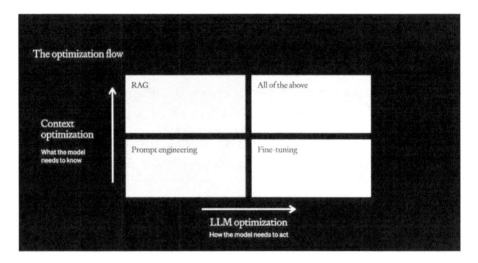

Figure 2-3. *Pattern used with generative AI*

Retrieval-Augmented Generation

Retrieval-augmented generation (RAG) is a pattern that combines the strengths of retrieval-based and generative approaches. In this technique, a retrieval model is used to identify and retrieve relevant information from a knowledge base or corpus, which is then provided as input to a generative language model. The generative model uses this retrieved information, along with the original input, to generate the final output. RAG is particularly useful for tasks that require synthesizing information from multiple sources, such as open-domain question answering. For example, when asked a question like "What are the main causes of climate change?", a RAG model could retrieve relevant passages from scientific articles and reports, and then use a generative model to produce a coherent and informative answer by synthesizing the retrieved information.

Fine-Tuning

Fine-tuning is a technique used to adapt a pre-trained generative model to a specific task or domain. It involves taking a model that has been pre-trained on a large, general corpus (e.g., a language model trained on a vast amount of text data), and then further training it on a smaller, task-specific dataset. This process allows the model to specialize and perform better on the target task while leveraging the knowledge and patterns learned during the initial pre-training phase. For instance, a large language model pre-trained on general text data could be fine-tuned on a dataset of product reviews to generate more accurate and relevant product descriptions or summaries.

Prompt Engineering

Prompt engineering is a crucial technique in the context of large language models and generative AI systems. It involves carefully crafting the input prompts or conditioning information provided to the model to guide its outputs in a desired direction. Effective prompt engineering can significantly improve the quality, coherence, and task-specific performance of the generated outputs. For example, when working with a language model for creative writing, providing a well-crafted prompt that sets the tone, genre, and context can lead to more engaging and relevant story generations. Prompt engineering often involves experimenting with different prompt structures, incorporating task-specific instructions, and leveraging techniques like few-shot learning to guide the model's generation process.

Model Deployment, Inference, and Optimization

Once your architecture is defined and customization is completed it's time to deploy the model. Deploying generative AI models is a very complex task but due to cloud Service providers (CSPs) like Amazon Web Services

(AWS), Google Cloud Platform (GCP), Microsoft Cloud Computing Service (Azure), Alibaba, Salesforce, Oracle, and IBM, and their capacity to spring off large compute resources, its relatively easy to deploy these generative AI models. Most of these cloud service providers have infrastructure to make these models available with API calls, which makes it cost-effective and easy to integrate.

To optimize generative AI model inference and deployment for efficiency, use deep learning containers and frameworks like SageMaker Inference containers, DeepSpeed, and Hugging Face Accelerate to implement techniques like pruning and quantization that reduce model size and memory usage. Deploy models on more energy and cost efficient instances like Inferential from AWS and Tensor Processing Unit from GCP which offer highly energy-efficient inference for large models, or use batch transform to avoid maintaining continuous infrastructure when real-time response is not needed.

Continuous Improvement

The generative AI lifecycle continues even after deployment. Every day, fresh data is generated. According to the latest estimates, ~328.77 million terabytes of data is created each day. The nature of generative AI models necessitates adding more data and retraining. Over time updates help protect performance in changing real-world contexts. Responsible development is driven by a continuous evaluation of the ethical and legal ramifications.

Unique Characteristics of the Generative AI Lifecycle

The lifecycle of generative AI models exhibits several unique characteristics that differentiate it from classic AI approaches. One of the most distinctive aspects is the iterative nature of training and fine-tuning. Unlike traditional AI models, which are typically trained once and occasionally updated, generative AI models often undergo continuous iterative training and fine-tuning cycles. This iterative process is crucial for improving the model's performance, adapting to new data distributions, and mitigating potential biases or safety concerns.

Another unique characteristic of the generative AI lifecycle is the significant computing requirements and the need for specialized hardware accelerators. Generative AI models, particularly large language models and diffusion models, can have billions or even trillions of parameters, making them computationally intensive during training and inference. This necessitates the use of high-performance computing resources, such as powerful GPUs and TPUs, to enable efficient training and deployment of these models.

Moreover, the data requirements for generative AI models are often more demanding compared to classic AI approaches. Generative AI models thrive on large, diverse, and high-quality datasets, which can include unstructured data sources like text corpora, images, audio, and video. Efficient data management, preprocessing, and curation pipelines are crucial to ensure the model's performance and mitigate potential biases or hallucinations.

The evaluation metrics used for generative AI models also differ from those employed in classic AI. While traditional AI models are often evaluated based on metrics like accuracy, precision, and recall, generative AI models require task-specific metrics such as perplexity, BLEU score, Fréchet Inception Distance (FID), and others. These metrics are designed to assess the quality, coherence, and diversity of the generated outputs, which can be challenging to quantify.

Challenges in the Generative AI Lifecycle

The generative AI lifecycle presents several unique challenges that must be addressed to ensure the responsible and effective development and deployment of these powerful models. The development and deployment of large-scale generative AI models pose significant environmental challenges due to their substantial computational requirements and energy consumption. Balancing the benefits of generative AI with its environmental costs remains a crucial challenge in the AI lifecycle, necessitating more energy-efficient training methods and carbon-aware machine learning practices. Here are some more challenges that we need to be aware of:

- Model collapse: A model can collapse when it is unable to learn the full diversity and complexity of the training data. When this occurs, the model overfits and becomes biased toward only a few simplified patterns in the data. For example, let's say you are training an image generation model to output photos of different people. If mode collapse happens, the model may start generating similar looking faces repeatedly, instead of a wide variety of faces.

- Model degeneration: A gradual loss of capabilities of the model over time during training can lead to degeneration. There can be multiple causes— unstable gradients, getting trapped in local optima, accumulating errors. For instance, the accuracy or ability of the model to generate coherent text/images may start deteriorating as training progresses. It essentially forgets or unlearns how to perform the task properly, due to issues in the modeling approach or training data/procedure.

- Ethical considerations and potential misuse: It is important to navigate the ethical considerations and potential misuse of these models. Generative AI models have the capability to generate highly realistic and convincing content, which raises concerns about potential biases, privacy violations, and the spread of misinformation or deep fakes. Addressing these ethical concerns requires robust governance frameworks, rigorous testing and monitoring, and the incorporation of ethical principles throughout the model development lifecycle.

- Interpretability and explainability: The interpretability and explainability of generative AI models pose significant challenges. Unlike traditional AI models, which can be relatively easy to interpret, the inner workings and decision-making processes of generative AI models are often opaque and difficult to understand. This lack of interpretability can hinder trust, accountability, and the ability to identify and mitigate potential biases or errors. Techniques such as feature attribution, saliency maps, and visualizations are being explored to improve the interpretability of generative AI models, but significant research and development efforts are still required in this area.

- Cybersecurity and fraud: The generative AI lifecycle also introduces new challenges related to cybersecurity and fraud. As these models become more powerful and ubiquitous, they may be exploited for malicious purposes, such as generating convincing phishing attacks, spreading misinformation, or engaging in financial fraud. Robust security measures, including adversarial training, watermarking, and advanced detection techniques, are necessary to mitigate these risks and ensure the safe and responsible deployment of generative AI technologies.

Optimization and Inefficiencies in Generative AI Lifecycle

One key aspect that is challenging to identify bottlenecks and inefficiencies in AI workloads. The following are some areas where further optimization would be warranted.

- Data bottlenecks: Inefficient data loading and preprocessing pipelines can create bottlenecks, especially when working with large datasets or complex data formats. Slow data transfer rates or inefficient data storage solutions can also lead to bottlenecks during training or inference.

- Model architecture inefficiencies: Suboptimal model architectures or inefficient implementations can lead to longer training times, higher resource consumption, or poor performance during inference. Examples of such inefficiencies include inefficient attention mechanisms, redundant computations, or inefficient use of parallelism.

- Computational bottlenecks: Insufficient computational resources, such as limited GPU memory or CPU cores, can create bottlenecks during training or inference, leading to longer processing times or resource contention. Additionally, inefficient use of available computational resources, such as underutilized GPUs or CPUs, can also lead to inefficiencies.

- Communication bottlenecks: In distributed training setups, inefficient communication protocols or network bottlenecks can slow down the training process and lead to inefficiencies. Examples of such bottlenecks include slow data transfer between nodes, inefficient gradient synchronization, or network congestion.

- Inference bottlenecks: Inefficient model deployment or serving strategies can create bottlenecks during inference, leading to high latency or low throughput. Examples of such bottlenecks include suboptimal model quantization or pruning, inefficient batching strategies, or resource contention during inference.

- Inefficient resource utilization: Underutilized or overprovisioned resources can lead to inefficiencies and higher costs. Examples of inefficient resource utilization include idle GPUs, overprovisioned CPU resources, or inefficient autoscaling strategies.

Measuring and Monitoring AI Workload Performance

Measuring and monitoring AI workload performance is crucial for ensuring efficient resource utilization, identifying bottlenecks, and optimizing the overall system.

The Need for Transparency and Visibility in AI Workload Management

Transparency and visibility are vital for efficient AI workload management, enabling organizations to debug, optimize, and improve AI systems while ensuring collaboration, compliance, and cost management. Logging and observability provide visibility into the inner workings of AI systems, enabling developers and operators to understand what is happening within the system and identify potential issues or inefficiencies.

Logging involves capturing and storing relevant information, such as system events, metrics, and diagnostic data, during the various stages of the AI workload lifecycle, including data preprocessing, model training, and inference.

Observability goes beyond logging by providing the ability to analyze and correlate logged data, enabling deeper insights into the system's behavior and performance. Proper logging and observability practices are essential for troubleshooting, performance optimization, and ensuring the reliability and scalability of AI workloads.

Techniques for Monitoring and Analyzing AI Workload Performance
Metrics Collection and Monitoring

Collecting and monitoring relevant metrics related to resource utilization (CPU, GPU, memory, network), model performance (accuracy, loss, convergence), and system health (latency, throughput, errors) is essential. Tools like Prometheus, Grafana, and cloud-native monitoring solutions (e.g., NVIDIA SMI, AWS CloudWatch, GCP Monitoring) can be utilized for collecting and visualizing these metrics.

Distributed Tracing

Distributed tracing involves tracking and analyzing the flow of requests or operations across different components of a distributed AI system. Tools like Jaeger, Zipkin, and cloud-native tracing solutions (e.g., AWS X-Ray, GCP Cloud Trace) can help identify bottlenecks, latencies, and performance issues in distributed AI workloads.

Logging and Log Analysis

Implementing structured logging practices to capture relevant information during different stages of the AI workload lifecycle is crucial. Log aggregation and analysis tools like Elasticsearch, Logstash, Kibana (ELK stack), or cloud-native logging solutions (e.g., AWS CloudWatch Logs, GCP Logging) can be used to centralize and analyze logs. Implementing log parsing and querying capabilities enables extracting insights and identifying patterns or anomalies in the logged data. For example, Amazon Opensearch can perform automatic anomaly detection on the incoming log information.

Profiling and Debugging Tools

Utilizing profiling tools like TensorFlow Profiler, PyTorch Profiler, or cloud-native profiling solutions (e.g., AWS SageMaker Debugger, GCP AI Platform Profiler) is necessary to analyze model performance, identify bottlenecks, and optimize resource utilization during training and inference. Debugging tools like TensorFlow Debugger or PyTorch Debugger can be used to inspect and debug model behavior, identify issues with data or model parameters, and ensure correct execution.

Alerting and Notification Systems

Setting up alerting and notification systems is essential to proactively detect and notify stakeholders of performance issues, resource constraints, or other anomalies in the AI workload. Integrating with existing monitoring and observability tools enables defining alert rules and notification channels (e.g., email, Slack, PagerDuty).

Summary

This chapter explains common architecture patterns used in generative AI, such as retrieval-augmented generation, fine-tuning, and prompt engineering, each serving specific purposes and offering distinct advantages. However, the generative AI lifecycle also presents unique challenges, including mode collapse, model degeneration, ethical considerations, interpretability issues, and cybersecurity risks. Addressing these challenges requires robust governance frameworks, rigorous testing and monitoring, and the incorporation of ethical principles throughout the model development lifecycle.

Identifying and mitigating bottlenecks and inefficiencies in data pipelines, model architectures, computations, communication, inference, and resource utilization are crucial to optimizing AI workloads. Techniques such as metrics collection, distributed tracing, logging and log analysis, profiling tools, and alerting systems play a vital role in monitoring and analyzing AI workload performance, enabling organizations to debug, optimize, and improve their AI systems while ensuring collaboration, compliance, and cost management.

The next chapter discusses how to maximize the power of your generative AI workloads through expert techniques for optimizing hardware utilization, intelligent scheduling, and cost-effective resource management.

CHAPTER 3

Hardware Optimization for Generative AI

Introduction

The field of generative artificial intelligence (AI) has emerged as one of the most exciting yet computationally demanding areas within AI research today. Generative AI models, such as large language models, text to image/video/audio models, possess the remarkable ability to generate human-like text, images, video, and more, leading to a wide range of beneficial applications. However, the training and operation of these complex neural network models require immense amounts of computing power, which poses a significant challenge.

As generative AI continues to advance at a rapid pace, the selection of hardware that optimizes both performance and energy efficiency has become a critical consideration. The latest generation of hardware, including graphics processing units, tensor processing units, and field-programmable gate arrays, offers new opportunities to run generative workloads more sustainably.

This chapter aims to analyze the importance of choosing energy-efficient hardware for generative AI and examine the trade-offs of available options. By delving into the intricacies of hardware selection for generative AI, we can better understand the factors that contribute to efficient and sustainable computing in this rapidly evolving field.

© Ishneet Kaur Dua and Parth Girish Patel 2024
I. K. Dua and P. G. Patel, *Optimizing Generative AI Workloads for Sustainability*,
https://doi.org/10.1007/979-8-8688-0917-0_3

The chapter explores the unique computational requirements of generative AI models and how different hardware architectures can address these demands. Furthermore, the chapter investigates the trade-offs between performance, energy efficiency, and cost, as these factors play a crucial role in the decision-making process for organizations and researchers working with generative AI.

The Rising Significance of Hardware Efficiency in Generative AI

In recent years, generative AI has experienced a surge in interest and investment, captivating the attention of tech giants and startups alike. The ability to automatically generate high-quality, realistic outputs such as images, videos, speech, and text has unlocked new possibilities in various fields. However, developing state-of-the-art generative models like OpenAI's GPT-3 language model and DeepMind's AlphaFold protein folding model relies on massive datasets and extensive computational resources.

The cost of training these models is staggering. For example, GPT-3 likely required around 3,640 petaflop-days of compute for training which can get very expensive. This restricts access to powerful generative models for smaller organizations and hinders innovation. Moreover, running inference on trained models at scale incurs high costs when relying solely on traditional hardware designed for peak compute performance. Figure 3-1 shows the sustainability impact of training the GPT-3 model.

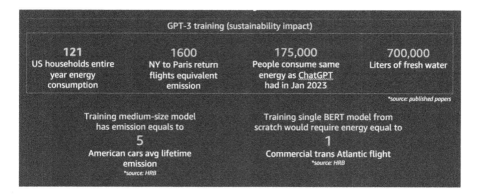

Figure 3-1. *In PricewaterhouseCoopers (PwC's) 2023 emerging tech survey, just 22% of business leaders cited sustainability impact as a top issue in GenAI deployment*

As generative models and datasets grow, effective hardware optimization, including specialized accelerators, is crucial for sustainable progress. Energy-efficient hardware can mitigate the environmental impact of developing and deploying generative AI while controlling costs, enabling a wider range of users to benefit from state-of-the-art models.

Major chip manufacturers have risen to the challenge, developing custom silicon optimized for the matrix multiplication and accumulation operations core to GenAI. These chips offer massive parallelism, high memory bandwidth, and novel architectures like in-memory computing to deliver superior performance per watt compared to traditional CPUs and GPUs. Companies like AMD, NVIDIA, Amazon, Intel, Apple, and startups like EnCharge AI are pioneering new AI-focused chips spanning a range of capabilities—from low-power neural processing units (NPUs) for on-device GenAI in smartphones to high-throughput training chips for multi-billion parameter foundation models in the cloud. Figure 3-2 outlines the key features, target use cases, and manufacturers driving innovation in this critical hardware frontier for next-generation AI.

Hardware	Key Features	Use Cases	Companies
CPU	Versatile, sequential processing	General-purpose computing	Intel, AMD, ARM
GPU	Highly parallel, CUDA programming	Deep learning training, rendering	NVIDIA
NPU	Dedicated neural network processing	AI inference in edge devices	AMD
TPU	Specialized for neural networks	Deep learning inference	Google
FPGA	Reconfigurable, low-latency inference	Custom AI acceleration	Xilinx, Intel
ASIC	Fixed-function, high-performance	Deep learning training, inference	Google, Apple, Graphcore
Neuromorphic	Mimics brain architecture, low power	Pattern recognition, edge computing	IBM, Intel

Figure 3-2.   Comparative analysis of hardware accelerators https://medium.com/@AbhibrotoM/accelerating-ai-understanding-the-hardware-behind-genai-ff38b3e175b0

Comparing Hardware Options for Generative AI

For the past decade, GPUs have been the driving force behind accelerating various applications, from traditional high-performance computing to cutting-edge deep learning workloads. Generative AI models have heavily relied on GPUs' massively parallel processing power during training and inference. Compared to alternative options, GPUs offer exceptional performance for mixed-precision workloads, which are common in deep learning, and provide high memory bandwidth to support data-intensive models.

Graphics Processing Units

NVIDIA's Volta and Ampere graphics processing units (GPUs) are highly efficient for generative AI, thanks to their tensor cores optimized for four-way mixed-precision operations and sparse linear algebra. For instance, models similar to GPT-3 can be trained up to 18 times faster on Ampere-based GPUs than older Pascal GPUs. This accelerated training enables researchers to iterate experiments more quickly and explore new possibilities. See Figure 3-3.

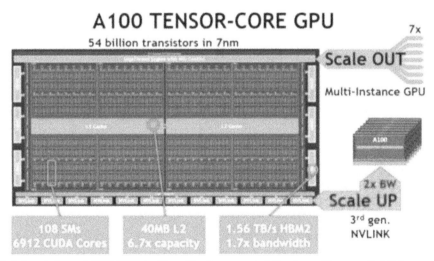

High-level view of A100 GPU, built on the Ampere architecture. (Source: Nvidia)

Figure 3-3. *High-level view of A100 GPU built on Ampere architecture*

Despite their advantages, GPUs consume significant energy, depending on factors like architecture, utilization, and precision needs. As a result, more efficient alternatives may be better suited for certain generative AI scenarios, especially at scale.

Tensor Processing Units

A tensor processing unit (TPU) is specialized hardware designed to accelerate neural networks for deep learning workloads. They offer extreme parallelism and higher performance per watt than conventional hardware, making them ideal for demanding applications. For instance, a study indicated that Google's fourth generation TPUv4 pods can deliver up to 1 exaflops of mixed-precision computing power while consuming only 20 megawatts of power.

TPUs' systolic array architecture and model execution units, optimized for low-precision operations, make them particularly well-suited for inference of trained generative models. Additionally, TPUs can scale training to thousands of chips through Google's TPU mesh topology, enabling the training of models with trillions of parameters.

While TPUs offer impressive performance and efficiency, they have limitations, including reduced numerical precision and memory compared to GPUs. Moreover, they are proprietary to Google and only available via its cloud platform. For smaller organizations and non-Google Cloud users, GPUs may remain a better option despite slightly lower efficiency.

Field-Programmable Gate Array

A field-programmable gate array (FPGA) is a highly adaptable hardware component that can be programmed on the fly to meet the specific needs of various applications. In the context of generative AI workloads, FPGAs offer reduced latency and optimized power usage in certain scenarios, such as recommendation systems, making them an attractive alternative to GPUs.

Microsoft, for instance, has developed a deep learning recommendation model inference stack on FPGAs, achieving a remarkable 58 times higher performance per watt compared to GPUs. This is made possible by features like aggressive floating-point compression and the ability to leverage lower bit-rate integers for quantization, which some generative networks require. See Figure 3-4.

Figure 3-4. *Field-Programmable Gate Array*

While FPGAs offer impressive flexibility and performance, their programming complexity remains a challenge. In contrast, GPUs and TPUs provide faster system development, iteration speed, and improved mixed-precision support, making them more suitable for training cutting-edge generative models. Ultimately, no single accelerator can meet all AI hardware needs. However, carefully selecting the right option for specific workload parameters can yield significant efficiency and performance gains over one-size-fits-all GPUs.

Application-Specific Integrated Circuits

An application-specific integrated circuit (ASIC) is a custom-built chip designed to deliver maximum performance and efficiency for specific applications, including AI. Tech giants like Google, Apple, and Graphcore are harnessing the potential of ASICs, exemplified by Google's TPU, Apple's Neural Engine, and Graphcore's IPU. These chips offer unparalleled performance and energy efficiency, making them ideal for deep learning tasks. However, designing and manufacturing ASICs requires significant upfront investment.

Neuromorphic Chips

Neuromorphic chips draw inspiration from the human brain's architecture, mimicking the behavior of neurons and synapses. They excel in tasks like pattern recognition and low-power sensor data processing, offering great potential for AI applications. Companies like IBM and Intel are at the forefront of neuromorphic computing research, with chips like IBM's TrueNorth and Intel's Loihi. Neuromorphic chips hold significant promise for low-power, real-time AI tasks, such as edge computing and autonomous systems.

Neural Processing Unit

A neural processing unit (NPU) for GenAI is a specialized hardware component designed to accelerate and optimize the performance of GenAI models on edge devices. NPUs are built to handle the unique computational requirements of GenAI, which involve complex matrix multiplications, nonlinear functions, and data movement.

The following are some key features of NPUs.

- Matrix multiplication engine: Efficient for transformer-based models like GenAI

- Nonlinear functions: Optimized using a flexible lookup table approach

- Memory hierarchy: Flexible memory management, including on-chip SRAM

- Dedicated DMA: Efficient and pipelined data flow

- Bandwidth reduction techniques: Maximizing bandwidth and minimizing bottlenecks

Let's summarize the advantages and limitations of the various options in the table below.

Table 3-1. *The Advantages and Limitations of Hardware Accelerators*

Hardware	Advantages	Limitations
GPUs	Excellent support for mixed precision and dense matrix calculations, scale well for large batch sizes	Not optimized for low-precision INT8/INT4 operations, higher latency due to CPU-GPU data transfer
TPUs	Extremely high teraFLOP/Watt efficiency for low-precision matrix math, tight coupling to models quantized for 8-bit ops	Constrained to Google Cloud ecosystem, limited numerical precision support
FPGAs	Flexible bit precision for quantized models, hardware-level pipeline optimization reduces serial compute latency	Programming complexity remains higher, often need HLS code translation

(continued)

65

Table 3-1. (*continued*)

Hardware	Advantages	Limitations
ASICs	Custom-built chips designed for specific applications, ideal for deep learning tasks	Require significant upfront investment
Neuromorphic Chips	Inspired by the human brain's architecture, excels in tasks like pattern recognition and low-power sensor data processing	Limited availability and support
NPUs	Specialized hardware components designed to accelerate and optimize the performance of generative AI models on edge devices	Limited availability and support

Key Considerations for Hardware Selection

The rapid advancement of deep learning and generative adversarial networks (GANs) has triggered an unprecedented surge in the size and computational requirements of AI models. The latest generative models, such as GPT-3, DALL-E 2, and AlphaFold, boast an exponential increase in parameters and datasets compared to their predecessors. However, this growth comes at a cost: training and operating these data and compute-intensive models demand a substantial increase in processing power.

As a result, organizations seeking to develop or utilize generative AI capabilities face a significant challenge: selecting specialized hardware that balances performance, efficiency, and scalability. The choice of accelerator can make or break the ability to train custom models within a reasonable budget and timeframe. Moreover, inferencing trained generative models at scale requires efficient and cost-effective hardware. Furthermore, minimizing environmental impact by reducing carbon emissions should be a crucial consideration during hardware evaluation.

Organizations must objectively evaluate various hardware options against key criteria to navigate this complex landscape. The following sections provide an in-depth, unbiased analysis of each factor, empowering organizations to make informed decisions about accelerators for their generative AI projects.

Compute Performance Requirements

The scale of computation necessary for state-of-the-art generative model training and inference continues rising dramatically fast. OpenAI's 2020 GPT-3 model required intense amounts of computing power to train, estimated to be approximately 356 GPU years, according to an OpenAI blog post. To put that into perspective, training GPT-3 would take 9,000 years on a single NVIDIA V100 GPU!

More specifically, OpenAI highlights that different-sized GPT-3 model variations are needed.

- 125 million parameter model: 1 GPU year

- 1.5 billion parameter model: 6 GPU years

- 6.7 billion parameter model: 76 GPU years

- 175 billion parameter model (GPT3): 356 GPU years

This exponential growth in computational demand comes from increasing the number of parameters in the transformer-based language model architecture to over 175 billion to ingest more text data and improve performance on natural language tasks. See Figure 3-5.

Model Name	n_{params}	n_{layers}	d_{model}	n_{heads}	d_{head}	Batch Size	Learning Rate
GPT-3 Small	125M	12	768	12	64	0.5M	6.0×10^{-4}
GPT-3 Medium	350M	24	1024	16	64	0.5M	3.0×10^{-4}
GPT-3 Large	760M	24	1536	16	96	0.5M	2.5×10^{-4}
GPT-3 XL	1.3B	24	2048	24	128	1M	2.0×10^{-4}
GPT-3 2.7B	2.7B	32	2560	32	80	1M	1.6×10^{-4}
GPT-3 6.7B	6.7B	32	4096	32	128	2M	1.2×10^{-4}
GPT-3 13B	13.0B	40	5140	40	128	2M	1.0×10^{-4}
GPT-3 175B or "GPT-3"	175.0B	96	12288	96	128	3.2M	0.6×10^{-4}

Figure 3-5. *Sizes, architectures, and learning hyper-parameters (batch size in tokens and learning rate) of the models that we trained. All models were trained for a total of 300 billion tokens. Source: OpenAI*

The sheer computing scale for modern state-of-the-art generative models like GPT-3 emphasizes why specialized hardware accelerators with high throughput are so important. Even advanced research labs struggle to keep up with the fast-rising computational budget required by ever-growing models.

Memory Bandwidth Needs

In addition to raw computing capacity measured in teraflops, delivering data to feed the hungry neural network beasts is equally crucial. Models with billions and trillions of parameters need to fetch weights and activate connections between layers at blistering speeds to fully utilize available computing power.

For example, a large transformer-based language model leverages self-attention—connecting each token to every other token from all positions. This requires immense memory bandwidth to query the activation of a single token across both model weights and intermediate activations of other tokens in parallel per layer. Unimpeded data flow is vital to accelerate matrix multiply-accumulate operations fundamental to neural network computations.

Architectures like NVIDIA's Hopper GPUs provide 3 TB/sec of memory bandwidth to eliminate data transfer bottlenecks during large batch training. Similarly, Google TPU v4 pods achieve 1.2 TB/sec per chip to feed models during massively parallel inference. Even higher bandwidth technologies like High Bandwidth Memory (HBM) integrated directly on accelerator die stacks promise to overcome memory walls as model scales continue trending upward without slowing innovation cycles. See Figure 3-6.

Figure 3-6. *NVIDIA Grasshopper architecture*

Carefully evaluating these memory subsystem characteristics helps prevent choosing hardware that artificially caps model size or dataset limitations due to insufficient memory bandwidth for target applications. The right balance of compute capacity and data feeding ability ensures selected accelerators fully exploit model parallelism for optimal speed and efficiency.

Numerical Precision Capabilities

While brute force flops (floating point operations per second) help, not all are equally effective. The numerical format used for calculations impacts results. Various formats, including FP32, FP16, BF16, TF32, and INT8, offer trade-offs between precision, speed, and memory usage. For instance, TF32, found in Ampere GPUs, balances precision and throughput, accelerating calculations up to 30 times faster than FP32. Understanding these formats and their implications for specific workloads is crucial for matching models with suitable hardware. Leading AI hardware vendors now natively accelerate multiprecision operations, recognizing the benefits of using formats beyond FP32 for certain generative modeling architectures.

Model Architecture Specifics

To choose hardware well-aligned with ecosystem needs, it's essential to carefully evaluate model architecture details beyond headline FLOP counts or memory capacity. Key details that inform accelerator matchmaking relate to tensor operations, sparsity, and emerging data types. Various vendors now optimize for dimensions such as sparse tensor compute, revealing 5x–7x speedups over previous generational hardware for models leveraging sparsity.

Tensor Operations

Tensor operations form the foundation of model architecture, and accelerators are designed to optimize these operations. Google's Tensor Processing Unit (TPU) exemplifies this, specifically designed to accelerate tensor-related calculations. TPU hardware is optimized for vector and matrix operations, which are fundamental to many AI applications.

Sparsity

Sparsity plays a vital role in model architecture, significantly influencing accelerator performance. By optimizing for sparse tensor compute, vendors have achieved remarkable 5x–7x speedups over previous hardware generations for models that leverage sparsity. This improvement stems from the reduced number of operations required, resulting in enhanced performance and energy efficiency.

Emerging Data Types

Emerging data types, such as those used in high-order FEM and tensor decomposition, are also critical in informing accelerator matchmaking. For example, the development of high-performance hardware architectures for tensor singular value decomposition (t-SVD) has shown significant speedups over software implementations on different t-SVD tasks. T-SVD is an extension of the traditional SVD for tensors (multidimensional arrays). This highlights the importance of considering emerging data types when choosing hardware for specific ecosystem needs.

In conclusion, thoroughly evaluating model architecture is essential for choosing hardware that aligns with ecosystem needs. Organizations can ensure optimal accelerator matchmaking by considering key factors like tensor operations, sparsity, and emerging data types, leading to improved performance and energy efficiency.

Training vs. Inference Priorities

While closely related to the production deployment of AI solutions, training, and inference entail different concerns for hardware selection. Batch size, latency, and throughput targets help determine suitable hardware. Specialized hardware accelerators can enhance inference efficiency in terms of cost and power consumption, but may sacrifice training capabilities. Cutting-edge model research relies on hardware that can handle large vocabularies, complex representations, and efficient sampling required for advancing generative AI models.

Training and Inference: Distinct Priorities for AI Hardware Selection

The A100 GPU offers significant performance improvements over the V100 for BERT pre-training and inference. For training, the A100 delivers up to 6x faster performance on FP32 and 3x on FP16, while for inference it provides up to 7x higher throughput compared to the V100. Performance comparison between V100 and A100 NVIDIA architectures on BERT pretraining throughput and inference can be seen in the following chart.

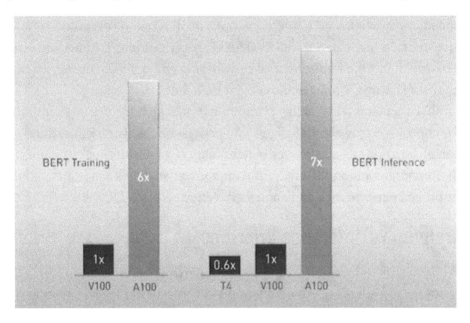

Figure 3-7. *Performance comparison between V100 and A100 NVIDIA architectures on BERT pre-training throughput and inference*

When deploying AI solutions in production, training, and inference are closely related yet distinct processes requiring different hardware selection considerations. A crucial question arises: Will the hardware be primarily used for training new generative models, efficient inference of available models, or a balance between both scenarios? The answer significantly influences the choice of hardware, as batch size, latency, and throughput targets play a critical role in determining suitable hardware.

Certain accelerators excel in optimizing inference cost and power efficiency, often at the expense of trainability. This is particularly important for scenarios where efficient inferencing is the primary concern. However, leading-edge model experimentation requires hardware that can support advanced capabilities, such as vocabulary capacity, representational complexity, and sample efficiency, necessary for breakthroughs in next-generation generative models.

Understanding the priorities of training and inference is vital for selecting the right hardware, as it directly impacts the performance and efficiency of AI applications. By recognizing these distinctions, organizations can make informed decisions to optimize their hardware selection and drive innovation in AI development.

Cloud Lock-in Constraints

The ease of leveraging cloud-hosted solutions has made public cloud providers an attractive option for organizations seeking readymade AI tooling. However, reliance on proprietary cloud hardware and services poses a significant risk of costly long-term lock-in. In contrast, portability between on-prem infrastructure and multi-cloud deployments offers greater flexibility in responding to evolving business objectives.

When selecting a hardware ecosystem, it's crucial to understand the external dependencies introduced by that choice. While cloud-managed services can accelerate velocity for smaller teams getting started with generative capabilities, custom models at scale, require careful evaluation of lock-in risks relative to hardware and cloud-agnostic options. Organizations must weigh the benefits of public cloud providers against the potential risks of lock-in, considering the long-term implications for their AI strategies. By prioritizing portability and understanding external dependencies, organizations can ensure greater flexibility and adaptability in their AI endeavors.

Power Budget and Form Factor Limitations

Generative models' insatiable appetite for computational resources and energy necessitates careful consideration of power consumption and real estate footprint during hardware evaluation. While supercomputer-scale setups are necessary for training large language models like Amazon Titan or Anthropic Claude family of models, most teams operate within more practical constraints.

Power Consumption

Power consumption significantly impacts operational costs, heat generation, and environmental sustainability. Ambarella aims to bring generative AI capabilities to edge devices while maintaining low power consumption, enabling cost-effective deployment. Their N1 SoC series supports multimodal AI applications, including large large language models (LLMs), with optimized power efficiency and a reduced form factor, making it possible to run generative AI models on edge devices.

Real Estate Footprint

The A100-based DGX system significantly reduces power consumption compared to its Volta-generation predecessor, requiring only 100kW instead of 630kW. This substantial decrease in power requirements also leads to a smaller physical footprint, making the A100 DGX more efficient and space-saving.

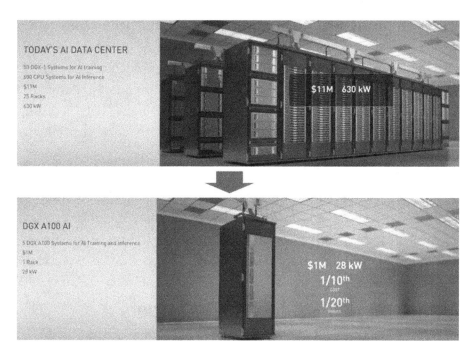

Figure 3-8. *Comparison of Volta-generation DGX (required 630kW of power) with A100 DGX (required A100kW of power) reduces power consumption and real estate footprint*

The physical space required for generative model training and deployment is another critical consideration. Amazon Web Services (AWS) offers Amazon EC2 P5 Instances (based on NVIDIA H100) designed for large-scale training workloads, reducing the real estate footprint required. Additionally, AWS provides Amazon EC2 Capacity Blocks for machine learning, allowing customers to reserve GPUs for future use, reducing on-premise infrastructure needs. NVIDIA shows that the same power can be achieved with their updated GPUs, which can complete the same task using one tenth of the speed and one twentieth of the power.

Most teams operate within practical power envelopes and physical space constraints, making evaluating hardware based on these considerations essential. Ambarella's N1 SoC series provides server-grade performance under 50W, suitable for various applications. Similarly, AWS's Amazon EC2 Inferentia 2 instances run high-performance deep-learning inference applications at scale globally while being the most cost-effective and energy-efficient option.

The power consumption and real estate footprint of generative models are critical considerations during hardware evaluation. By evaluating hardware based on these considerations, teams can ensure that they operate within practical power envelopes and physical space constraints while still achieving the performance required for generative AI workloads.

Environmental Sustainability Targets

With skyrocketing computational demands from areas like generative AI, optimizing for energy efficiency serves both ethical and business motivations. More computationally powerful hardware tends to require increased data center resources, driving energy usage higher. At certain thresholds, efficiency improves again through economies of scale. The environmental impacts of large language models contain both risks and opportunities. Widespread adoption of these systems could directly or indirectly affect the environment in positive or negative ways. Additionally, the capabilities they unlock may transform how environmental research is conducted. Multiple factors regarding their development and deployment influence the net sustainability outcome of harnessing these models.

As models continue to grow in size and complexity, their environmental impact will only increase unless proactive measures are taken. See Figure 3-9.

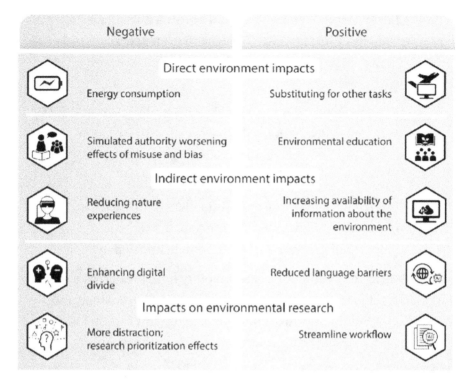

Figure 3-9. *Generative AI impact on sustainability* https://www.
researchgate.net/figure/Large-language-models-come-with-
risks-and-opportunities-for-the-environment-Increased_
fig1_368755796

To address the environmental challenges posed by generative AI,
several strategies and best practices should be considered.

- Optimize model architectures and training processes
  for energy efficiency. This can involve techniques like
  model pruning, quantization, and distillation, which
  can reduce computational requirements without
  significantly compromising performance.

- Leverage more energy-efficient hardware accelerators, such as TPUs and ASICs, which are designed to deliver high performance while minimizing power consumption.

- Explore techniques like model compression and knowledge distillation, which can reduce the memory and computational requirements of large models, thereby reducing their environmental impact.

- Implement carbon offsetting initiatives or invest in renewable energy sources to mitigate the carbon footprint of generative AI development and deployment.

Considering environmental sustainability from the early stages of model development and deployment is crucial. Researchers and organizations should prioritize energy efficiency and carbon footprint reduction as key design criteria, alongside performance and accuracy metrics.

Addressing the environmental challenges of generative AI requires collaboration and knowledge-sharing among researchers, organizations, and policymakers. By working together and adopting best practices, the AI community can harness the potential of generative AI while minimizing its environmental impact and promoting sustainable development.

Optimizing Hardware Utilization and Scheduling

Optimizing hardware utilization and scheduling is crucial for efficient and cost-effective deployment of GenAI workloads. GenAI models require significant computational resources, and suboptimal hardware utilization can lead to increased costs, longer training times, and reduced model performance. Next, let's discuss techniques for optimizing hardware utilization and scheduling in GenAI workloads, as well as strategies for efficient resource allocation and load balancing.

GPU Utilization

GenAI models are typically trained on GPUs, which are designed to handle massive parallel processing. Optimizing GPU utilization is critical to accelerate training times and reduce costs. Techniques include the following.

- GPU pinning: AWS provides GPU pinning through its Elastic Compute Cloud (EC2) instances, allowing you to assign specific GPUs to specific tasks.

- GPU virtualization: AWS offers GPU virtualization through its NVIDIA GPU Cloud and Amazon Elastic Container Service for Kubernetes (EKS), enabling multiple tasks to share the same physical GPU.

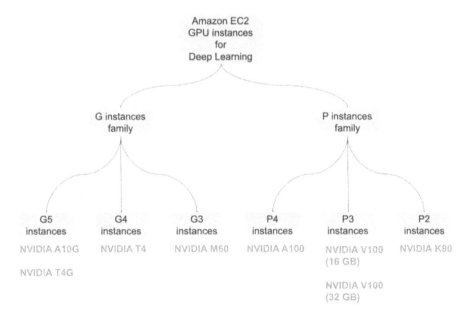

Figure 3-10. *Amazon EC2 GPU instances for deep learning*

CPU Utilization

While GPUs are the primary processing units for GenAI, CPUs are still required for data preprocessing, model serving, model inference and other tasks. Optimizing CPU utilization involves the following.

- CPU pinning: AWS provides CPU pinning through its EC2 instances, allowing you to assign specific CPUs to specific tasks.

- CPU virtualization: AWS offers CPU virtualization through its EC2 instances and EKS, enabling multiple tasks to share the same physical CPU.

Memory Optimization

GenAI models require significant memory to store model parameters, input data, and intermediate results. Optimizing memory utilization involves the following.

- Memory profiling: AWS and other cloud providers provides memory profiling tools, such as NVIDIA SMI, Amazon CloudWatch and AWS X-Ray, to identify memory bottlenecks and optimize memory allocation.

- Memory compression: AWS offers memory compression through its Amazon Elastic Block Store (EBS) and Amazon Simple Storage Service (S3), reducing memory usage and improving data transfer efficiency.

Storage Optimization

GenAI models require significant storage for model checkpoints, input data, and output results. Optimizing storage utilization involves the following.

- Storage tiering: AWS provides storage tiering through its S3, allowing you to use different storage tiers (e.g., SSD, HDD, and object storage) to optimize storage costs and performance.

- Data compression: AWS offers data compression through its S3 and Amazon Elastic File System (EFS), reducing storage usage and improving data transfer efficiency.

Job Scheduling

Scheduling GenAI jobs involves allocating resources (e.g., GPUs, CPUs, and memory) to specific tasks. The following are some of the techniques.

- First-come, first-served (FCFS): Scheduling jobs in the order they are received can lead to resource underutilization and increased wait times.

- Shortest job first (SJF): Scheduling the shortest job first to minimize wait times and improve resource utilization.

- Priority scheduling: Scheduling jobs based on priority can improve resource utilization and reduce wait times for high-priority jobs.

- Resource allocation: Allocating resources to GenAI jobs involves the following.

- Static allocation: Allocating fixed resources to each job can lead to resource underutilization and increased costs.

- Dynamic allocation: Allocating resources dynamically based on job requirements can improve resource utilization and reduce costs.

- Load balancing: Load balancing involves distributing workload across multiple resources to improve resource utilization and reduce wait times. The following describes some of the techniques.

  - Round-robin scheduling: Distributing workload across multiple resources in a circular order can improve resource utilization and reduce wait times.

  - Least connection scheduling: Distributing workload across multiple resources based on the number of active connections can improve resource utilization and reduce wait times.

Strategies for Efficient Resource Allocation and Load Balancing

- Resource pooling: Pooling resources (e.g., GPUs, CPUs, and memory) to enable dynamic allocation and improve resource utilization. This is a feature that Amazon Sagemaker (end to end ML service) provides.

- Containerization: AWS provides containerization through its Elastic Container Service (ECS) and Elastic Kubernetes Service (EKS) services, enabling efficient resource allocation and improving resource utilization.

- Orchestration: AWS offers orchestration through its EKS and Amazon ECS, managing resource allocation, scheduling, and load balancing for GenAI workloads.

- Autoscaling: AWS provides autoscaling through its EC2 and EKS, scaling resources up or down based on workload demand to improve resource utilization and reduce costs.

- Monitoring and analytics: AWS offers monitoring and analytics through its CloudWatch, Sagemaker Profiler/ Debugger and X-Ray, providing insights into GenAI workload performance and identifying optimization opportunities.

Similar Services

The following are similar services available from other cloud service providers. See Figure 3-11.

- Google Cloud Platform (GCP) offers GPU acceleration, CPU optimization, memory optimization, and storage optimization through its Compute Engine, Kubernetes Engine, and Cloud Storage services.

- Microsoft Azure provides GPU acceleration, CPU optimization, memory optimization, and storage optimization through its Virtual Machines, Kubernetes Service, and Blob Storage services.

- Alibaba Cloud offers GPU acceleration, CPU optimization, memory optimization, and storage optimization through its Elastic Compute Service, Container Service, and Object Storage Service.

Figure 3-11. *Generative AI offerings from major cloud service providers*

By leveraging these techniques and services, you can optimize hardware utilization and scheduling for GenAI workloads on AWS and other CSPs, improving resource utilization, reducing costs, and accelerating training times.

Summary

Optimizing hardware utilization and scheduling is critical for efficient and cost-effective deployment of GenAI workloads. Techniques such as GPU utilization, CPU utilization, memory optimization, and storage optimization can improve resource utilization and reduce costs. Scheduling techniques such as job scheduling, resource allocation, and load balancing can improve resource utilization and reduce wait times. Strategies such as resource pooling, autoscaling, containerization, orchestration, monitoring, and analytics can improve resource utilization and reduce costs. Organizations can accelerate GenAI development, improve model performance, and reduce costs by implementing these techniques and strategies.

CHAPTER 4

Software Optimization for Generative AI

As generative artificial intelligence (AI) models continue to grow in size and complexity, the computational demands and environmental impact of training and deploying these models have become significant concerns. Large language models and other generative AI systems can consume enormous amounts of energy and contribute substantially to carbon emissions. To address these challenges, optimizing the software that powers generative AI is crucial for achieving sustainability and reducing the environmental footprint of these technologies.

This chapter explores various software optimization techniques and strategies for sustainable generative AI. We dive into model architecture optimization, efficient training algorithms, optimized software frameworks and libraries, efficient inference strategies, software optimization for energy efficiency, and green software engineering practices. Additionally, we examine real-world case studies and best practices from industry and research and discuss emerging trends and future directions in sustainable AI software.

© Ishneet Kaur Dua and Parth Girish Patel 2024
I. K. Dua and P. G. Patel, *Optimizing Generative AI Workloads for Sustainability*,
https://doi.org/10.1007/979-8-8688-0917-0_4

The Importance of Software Optimization for Sustainable Generative AI

Generative AI models, particularly large language models and image generation models, have demonstrated remarkable capabilities in tasks such as text generation, code generation, and image synthesis. However, the training and deployment of these models come at a significant computational cost, leading to high energy consumption and carbon emissions.

The environmental impact of AI has become a growing concern, with studies estimating that the carbon footprint of training a single large language model can be equivalent to the lifetime emissions of thousands of cars. As these models increase in size and complexity, their energy demands and environmental impact will only escalate further.

Software optimization plays a crucial role in mitigating the environmental impact of generative AI by improving computational efficiency, reducing energy consumption, and enabling more sustainable practices. By optimizing the software that powers these models, we can unlock significant energy savings, lower carbon emissions, and pave the way for more environmentally responsible AI development and deployment.

Computational Challenges and Environmental Impact of Generative AI Models

Generative AI models, particularly large language models and image generation models, pose significant computational challenges due to their immense size and complexity. These models often consist of billions or even trillions of parameters, requiring vast amounts of computational resources for training and inference.

Model Size and Computational Demands

The computational demands of generative AI models stem from several factors.

- Model size: Large language models and image generation models can have billions or trillions of parameters, leading to substantial memory requirements and computational overhead during training and inference.

- Training data: Training these models often requires massive datasets, which can be computationally intensive to process and store.

- Iterative training: Generative AI models typically undergo iterative training processes involving multiple epochs and fine-tuning stages, further increasing computational demands.

- Inference complexity: Generating high-quality outputs, such as long-form text or high-resolution images, can be computationally expensive during inference.

Environmental Impact

The computational demands of generative AI models translate into significant energy consumption and carbon emissions. See Figure 4-1.

Figure 4-1. *Carbon emissions*

The training and deployment of these models can have a substantial environmental impact due to the following factors.

- Energy consumption: The vast computational resources required for training and inference consume substantial amounts of energy, often relying on energy-intensive data centers (the energy required for cooling data centers, which is a significant contributor to the overall energy consumption) and high-performance computing (HPC) systems.

- Carbon emissions: The energy consumption associated with generative AI models contributes to greenhouse gas emissions and carbon footprints, exacerbating the effects of climate change.

- Resource depletion: The production and disposal of hardware components used in AI systems can contribute to resource depletion and electronic waste.

- Scalability challenges: As generative AI models grow in size and complexity, their environmental impact will escalate further, posing significant sustainability challenges.

Software optimization techniques and strategies are crucial to address these computational challenges and mitigate the environmental impact of generative AI. By optimizing the software that powers these models, we can improve computational efficiency, reduce energy consumption, and enable more sustainable practices in AI development and deployment.

Model Architecture Optimization

Optimizing the architecture of generative AI models is a critical step toward improving computational efficiency and reducing the environmental impact of these systems. This section explores various optimization techniques and strategies for model architecture, including efficient model design, parameter efficiency, architectural innovations, and model compression techniques.

Efficient Model Design

Designing efficient model architectures is crucial for reducing computational demands and improving overall performance. Several approaches can be employed to achieve efficient model design.

- Parameter efficiency: Explore techniques to reduce the number of parameters in the model while maintaining or improving performance. This can include techniques such as weight sharing, sparse connections, and efficient attention mechanisms.

- Architectural innovations: Explore novel architectural designs and variants of existing architectures, such as transformer variants and sparse attention mechanisms, to improve computational efficiency and reduce memory requirements.

- Model scaling strategies: Investigate efficient strategies for scaling models, such as model parallelism, data parallelism, and pipeline parallelism, to distribute computational workloads and leverage available hardware resources effectively.

- Conditional computation: Implement techniques that enable dynamic and conditional computation that allow the model to selectively activate or deactivate components based on input data, reducing unnecessary computations. These techniques can drastically help reduce the carbon impact during the inference stages.

- Modular and composable architectures: Design modular and composable architectures that allow for efficient reuse and combination of model components, enabling more efficient model development and deployment.

Model Compression Techniques

Model compression techniques aim to reduce the memory footprint and computational requirements of generative AI models while preserving their performance. Several compression techniques can be employed, as seen in Figure 4-2.

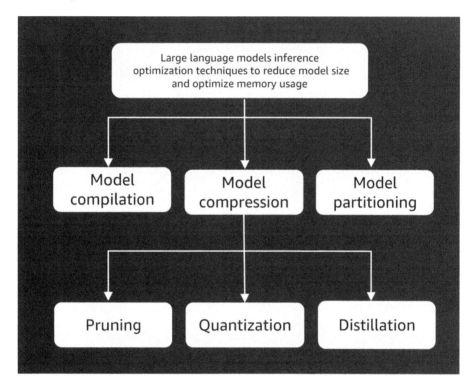

Figure 4-2. *Model compression techniques*

- Pruning: Identifying and removing redundant or less important parameters from the model effectively reduces its size and computational demands.

- Quantization: Reduce the precision of model parameters from floating-point to lower-precision representations, such as fixed-point or integer formats, without significantly impacting performance.

- Knowledge distillation: Transferring knowledge from a larger, more complex model (teacher) to a smaller, more efficient model (student) enables the deployment of compact models with comparable performance.

- Low-rank factorization: Approximating weight matrixes in the model using low-rank factorizations reduces the number of parameters and computational requirements.

- Sparsity techniques: Exploit sparsity in model parameters and activations to reduce memory requirements and enable efficient computations.

Combining efficient model design principles with model compression techniques makes it possible to develop generative AI models that are computationally efficient while maintaining high performance, thereby reducing their environmental impact.

Model Parallelism Strategies

Parallelizing the training and inference of generative AI models across multiple computational resources is essential for handling their immense computational demands. See Figure 4-3.

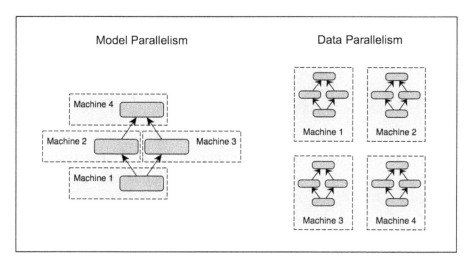

Figure 4-3. *Model parallelism techniques*

Several parallelism strategies can be employed.

- Data parallelism: Distributing the training data across multiple devices (e.g., GPUs or TPUs), allowing each device to process a subset of the data in parallel and synchronize gradients during training.

- Model parallelism: Partitioning the model across multiple devices, with each device responsible for a subset of the model's layers or components, enables processing larger models that would not fit on a single device.

- Pipeline parallelism: Dividing the model into sequential stages or partitions, with each stage running on a different device, enabling efficient utilization of available hardware resources and overlapping computations across stages.

- Hybrid parallelism: Combining multiple parallelism strategies, such as data parallelism and model parallelism, to leverage the strengths of each approach and optimize for different hardware configurations and model architectures.

Effective parallelism strategies can significantly reduce training and inference times, enabling more efficient utilization of computational resources and potentially reducing energy consumption and carbon emissions.

Efficient Training Algorithms

Optimizing the training algorithms and techniques for generative AI models can lead to significant computational savings and improved efficiency. Several approaches can be employed (see Figure 4-4).

Figure 4-4. *Training algorithm*

- Optimizers: Exploring efficient optimization algorithms, such as AdamW (weight decay), RAdam (Rectified Adam), and other variants of the Adam optimizer, which can converge faster and require fewer training iterations, reducing computational demands.

- Mixed precision training: Leveraging mixed-precision arithmetic, where computations are performed in lower precision (e.g., half-precision or BF16-bfloat16) while maintaining high-precision storage for model parameters, reducing memory requirements and computational overhead.

- Gradient checkpointing: Recomputing activations during the backward pass instead of storing them during the forward pass, reducing memory requirements at the cost of additional computations.

- Activation recomputation: Recomputing activations during the forward pass instead of storing them, trading off computational overhead for reduced memory requirements.

- Curriculum learning: Employing curriculum learning strategies, where the model is trained on simpler examples or tasks before gradually increasing the complexity, potentially improving convergence and reducing the overall computational cost of training.

Combining efficient training algorithms with model architecture optimization techniques makes it possible to achieve significant computational savings and improve the overall efficiency of generative AI model training.

Software Frameworks and Libraries

Optimized software frameworks and libraries enable efficient development, training, and deployment of generative AI models. This section explores various frameworks and libraries designed for efficient deep learning and generative AI and libraries for automatic mixed precision, quantization, and distributed training.

Efficient Deep Learning Frameworks

Deep learning frameworks, such as PyTorch and TensorFlow, provide the foundational infrastructure for developing and training generative AI models. See Figure 4-5.

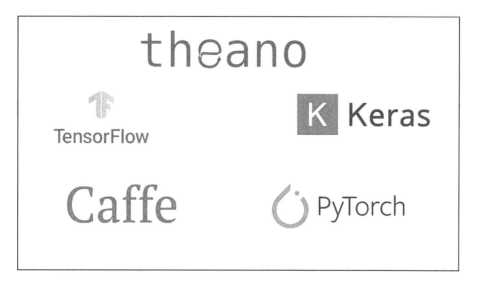

Figure 4-5. *Training algorithm*

These frameworks have been optimized for efficient computations, leveraging hardware accelerators like GPUs and TPUs and enabling parallelism and distributed training.

- PyTorch is a popular deep learning framework known for its dynamic computation graph and strong support for research and rapid prototyping. It provides efficient implementations of common operations, optimized for various hardware platforms, and supports features like automatic mixed precision and distributed training.

- TensorFlow is a widely-used deep learning framework developed by Google, offering a flexible and scalable ecosystem for building and deploying machine learning models. It supports efficient execution on various hardware platforms, including GPUs, TPUs, and CPUs, and provides tools for distributed training and model optimization.

- JAX (Just After eXecution) is a research-oriented framework for high-performance machine learning designed to be composable and differentiable. It leverages XLA (Accelerated Linear Algebra) for efficient execution on various hardware platforms, including GPUs and TPUs, and supports features like automatic batching and parallelization.

- ONNX (Open Neural Network Exchange) Runtime is a cross-platform, high-performance inference engine for deploying and running machine learning models. It supports efficient execution on various hardware platforms, including CPUs, GPUs, and specialized accelerators, and optimizes model quantization and pruning.

These deep learning frameworks provide the foundation for efficient development, training, and deployment of generative AI models, enabling researchers and developers to leverage hardware acceleration, parallelism, and optimized computations.

Optimized Libraries for Generative AI

In addition to general-purpose deep learning frameworks, several libraries have been developed specifically for efficient training and deployment of generative AI models, particularly large language models and image generation models.

- The Hugging Face Accelerate and Transformers library provides a comprehensive set of pre-trained models and utilities for natural language processing tasks, including language modeling, text generation, image analysis, and translation. It supports efficient training and inference on various hardware platforms and optimizes model parallelism and mixed precision training.

- Fairseq is a sequence modeling toolkit developed by Facebook AI Research, designed for efficient training and deployment of large-scale language models and other sequence-to-sequence models. It supports mixed precision training, gradient checkpointing, and distributed training across multiple GPUs and machines. See Figure 4-6.

Figure 4-6. *Training algorithm*

- NVIDIA NeMo (neural modules) is a conversational AI toolkit that provides building blocks for constructing and optimizing speech and language models. It supports efficient training and inference on NVIDIA GPUs, leveraging optimizations like mixed precision, tensor cores, and multi-GPU parallelism. See Figure 4-7.

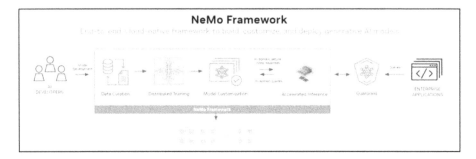

Figure 4-7. *Training algorithm*

- BigGAN-PyTorch is a library for efficient training and generating high-resolution images using Generative Adversarial Networks (GANs). It optimizes various GAN architectures and supports features like multi-GPU training and mixed precision computations.

These specialized libraries leverage optimizations tailored for generative AI models, enabling efficient training, inference, and deployment on various hardware platforms while providing tools and utilities for model parallelism, mixed precision, and other performance-enhancing techniques.

Automatic Mixed Precision and Quantization Libraries

Mixed precision training and quantization are essential for improving computational efficiency and reducing memory requirements in generative AI models. By using higher precision for the model's weights and lower precision for calculations, mixed precision training reduces memory usage and speeds up computations, especially on specialized hardware. This approach maintains the model's accuracy while making the training process more efficient. Another technique, quantization, simplifies the model's weights and calculations, making the model smaller and faster for predictions. Quantization can be applied after training or incorporated into the training process, allowing the model to adapt to the simplified representations. While quantization brings significant efficiency gains, it may slightly reduce the model's accuracy.

Several libraries have been developed to facilitate the adoption of these techniques.

- NVIDIA Apex is a PyTorch extension that provides tools for mixed precision training, enabling lower-precision data formats (e.g., FP16) for computations while maintaining high-precision storage for model parameters. It also includes utilities for distributed training and optimization techniques like gradient clipping and weight decay.

- TensorFlow Model Optimization Toolkit provides tools and APIs for quantization, pruning, and other model optimization techniques. It supports quantization-aware training, post-training quantization, and quantization-aware inference, enabling efficient deployment of quantized models on various hardware platforms.

- ONNX Runtime provides tools and utilities for quantizing and optimizing ONNX models for efficient inference. It supports various quantization techniques, including static quantization, dynamic quantization, and quantization-aware training, enabling efficient deployment on a wide range of hardware platforms.

- Hugging Face Optimum is a library for optimizing and deploying Hugging Face Transformers models, providing tools for quantization, pruning, and other optimization techniques. It supports efficient inference on various hardware platforms, including CPUs, GPUs, and specialized accelerators.

These libraries simplify the adoption of mixed precision training and quantization techniques, enabling researchers and developers to leverage the computational benefits of lower-precision arithmetic and optimized model representations while maintaining model performance.

Distributed Training Libraries and Frameworks

Distributed training is essential for efficiently training large-scale generative AI models, leveraging multiple computational resources and enabling parallelism. Several libraries and frameworks have been developed to facilitate distributed training.

- PyTorch Distributed is a module within PyTorch that provides APIs and utilities for distributed training across multiple GPUs and machines. It supports various communication backends, including Gloo, NCCL, and MPI, and enables data parallelism, model parallelism, and hybrid parallelism strategies.

- TensorFlow Distributed is a set of tools and APIs within TensorFlow for distributed training and deployment of machine learning models. It supports various distribution strategies, including data parallelism, model parallelism, and parameter server architectures, enabling efficient training on multiple GPUs, TPUs, and CPUs.

- Horovod is a distributed training framework that provides a simple and efficient way to leverage multiple GPUs and machines for training deep learning models. It supports various deep learning frameworks, including PyTorch, TensorFlow, and Apache MXNet, and enables data parallelism and model parallelism strategies.

- Ray is a distributed computing framework that
  provides a flexible and scalable platform for building
  and deploying distributed applications, including
  distributed training of machine learning models. It
  supports various parallelism strategies and enables
  efficient utilization of heterogeneous hardware
  resources.

These distributed training libraries and frameworks enable efficient
utilization of multiple computational resources, enabling parallelism and
scalability for training large-scale generative AI models. They provide tools
and utilities for data parallelism, model parallelism, and other distribution
strategies, facilitating efficient training and reducing overall computational
demands.

Efficient Inference Strategies

While training generative AI models is computationally intensive, efficient
inference strategies are crucial for enabling sustainable deployment
and real-time applications. This section explores various techniques
and strategies for optimizing inference performance, including model
quantization and pruning, optimized inference engines and runtimes,
batching and caching techniques, dynamic batching, and efficient serving
strategies.

Optimized Inference Engines and Runtimes

Efficient inference engines and runtimes are essential for deploying
generative AI models in production environments, enabling real-time
inference and low-latency responses. Several optimized inference engines
and runtimes have been developed to address the computational demands
of generative AI models. See Table 4-1.

Table 4-1. *Optimized Inference Engines and Runtimes for Generative AI Models*

Engine/ Runtime	Key Features	Supported Platforms	Optimization Techniques
TensorRT	- High-performance inference optimizer - Low-latency inference	NVIDIA GPUs	- Layer and tensor fusion - Kernel autotuning - Dynamic tensor memory management
ONNX Runtime	- Cross-platform inference engine - Supports various hardware	CPUs, GPUs, specialized accelerators	- Model quantization - Pruning - Hardware-specific optimizations
TensorFlow Lite	- Lightweight solution for mobile/ IoT - Optimized for edge devices	CPUs, GPUs, Edge TPU	- Optimized kernels and operators - Model compression
Apache TVM	- Machine learning compiler and runtime - Automatic optimization	CPUs, GPUs, specialized accelerators	- Operator fusion - Data layout transformation - Kernel autotuning
NVIDIA Triton Inference Server	- Cloud-native inference serving - Scalable deployment	NVIDIA GPUs	- Batching - Model ensembling - Dynamic batching

These optimized inference engines and runtimes leverage various optimization techniques, such as kernel fusion, autotuning, and specialized hardware acceleration, to enable efficient and low-latency inference for generative AI models. They are designed to optimize performance, reduce computational overhead, and facilitate efficient deployment in various environments, from cloud and data center deployments to edge and mobile devices.

Batching and Caching Techniques

Batching and caching techniques can significantly improve inference performance and efficiency for generative AI models by leveraging parallelism and reducing redundant computations.

- Batching involves processing multiple input samples as a batch, enabling parallelism and efficiently utilizing hardware resources. By processing batches of inputs together, the computational overhead can be amortized across multiple samples, leading to improved throughput and reduced latency.

- Caching involves storing and reusing intermediate computations or outputs to avoid redundant calculations. For generative AI models, caching can be applied at various levels, such as caching embeddings, attention matrixes, or even entire output sequences, reducing computational demands and improving inference efficiency.

- Attention caching in transformer-based models involves storing and reusing the attention matrixes computed during the self-attention mechanism. This can significantly reduce computational overhead, especially for tasks like text generation, where the attention matrixes can be reused across multiple decoding steps.

- Output caching in certain applications, such as question-answering or text generation, can be beneficial. By storing and retrieving previously generated outputs, redundant computations can be avoided, improving inference efficiency and reducing computational demands.

- Adaptive caching strategies can dynamically determine the optimal caching granularity and eviction policies based on factors like available memory, input characteristics, and computational requirements, ensuring efficient utilization of caching resources.

Batching and caching techniques can be combined with other optimization strategies, such as model quantization and pruning, to further enhance inference efficiency and reduce computational demands, enabling more sustainable deployment of generative AI models in various environments.

Dynamic Batching and Adaptive Batch Sizing

While batching can improve inference efficiency, static batch sizes may not be optimal for all scenarios, especially in environments with varying input loads or resource constraints. Dynamic batching and adaptive batch sizing techniques can address these challenges by dynamically adjusting batch sizes based on runtime conditions and resource availability.

- Dynamic batching involves dynamically forming batches of input samples based on their arrival rate and available computational resources. This approach can improve resource utilization and reduce latency by avoiding under-utilization or over-commitment.

- Adaptive batch sizing techniques dynamically adjust the batch size based on factors such as input characteristics, available memory, and computational resources. Balancing trade-offs between throughput, latency, and memory consumption can optimize performance and resource utilization.

- Load-aware batching strategies consider the current system load and resource availability when determining batch sizes. This can help maintain consistent performance and avoid resource contention or overloading in high-load scenarios.

- Heterogeneous batching techniques in scenarios with diverse input sizes or complexities can be employed to group inputs with similar computational requirements together, enabling more efficient utilization of resources and reducing performance variability.

- Batching policies and heuristics can be employed to determine optimal batch sizes and batching strategies based on input arrival rates, resource availability, latency requirements, and performance targets.

Dynamic batching and adaptive batch sizing techniques can significantly improve inference efficiency and resource utilization for generative AI models. This enables more sustainable deployment in dynamic and resource-constrained environments, such as cloud and edge computing scenarios.

Efficient Serving Strategies

Efficient serving strategies are crucial for deploying generative AI models in production environments, ensuring scalability, low latency, and efficient resource utilization. Several serving strategies and architectures can be employed.

- Serverless computing platforms, such as AWS Lambda, Google Cloud Functions, and Azure Functions, enable efficient and scalable workload deployment without the need for provisioning and managing servers. These platforms automatically scale resources based on demand, reducing idle resource consumption and enabling cost-effective deployment. Due to hardware sharing limitations, serverless computing is not used as mainstream for generative AI models, butt this is the area in which more innovation is expected.

- Containerization technologies, such as Docker and Kubernetes, facilitate efficient packaging, deployment, and scaling of generative AI models. Containers encapsulate the model and its dependencies, enabling portability and consistent execution across different environments, while Kubernetes provides orchestration and scaling capabilities for efficient resource management. AWS services such as Elastic Container Service (ECS) and Elastic Kubernetes Service (EKS) are managed services offering container orchestration environments

- Microservices architecture can improve the scalability and efficiency of generative AI model deployment. By decomposing the application into modular services, each responsible for specific tasks (e.g., text generation, image generation, or post-processing), resources can be allocated and scaled independently based on demand. This recommendation comes from 12-factor application framework which is designed around building optimized microservices for cloud native deployments

- Load balancing and autoscaling techniques can ensure efficient resource utilization and scalability for generative AI model deployments. Load balancers distribute incoming requests across multiple instances, while autoscaling mechanisms automatically adjust the number of instances based on demand, ensuring optimal resource allocation and minimizing idle resource consumption.

- Caching and content delivery networks (CDNs) can improve the efficiency and performance of generative AI model deployments by reducing redundant computations and minimizing network latency. Caching can store and serve frequently accessed outputs, while CDNs distribute content closer to end users, reducing network latency and improving overall responsiveness.

These efficient serving strategies leverage cloud-native technologies, containerization, microservices architectures, and scalable infrastructure to enable efficient and sustainable deployment of generative AI models in production environments, ensuring optimal resource utilization, scalability, and low latency.

Software Optimization for Energy Efficiency

In addition to optimizing computational efficiency, software optimization techniques can also target energy efficiency, reducing the overall energy consumption and carbon footprint of generative AI systems. This section explores various strategies for energy-efficient software optimization, including energy-aware scheduling and resource management, power-efficient software techniques, optimizing for heterogeneous hardware architectures, and leveraging specialized hardware accelerators.

Energy-Aware Scheduling and Resource Management

Energy-aware scheduling and resource management techniques aim to optimize the allocation and utilization of computational resources based on energy consumption and performance requirements, enabling more energy-efficient execution of generative AI workloads.

- Energy-aware task scheduling across available computational resources should consider energy consumption and performance trade-offs. This can involve techniques like consolidating workloads on fewer resources, migrating tasks to more energy-efficient resources, or dynamically adjusting resource allocation based on energy and performance requirements.

- Dynamic voltage and frequency scaling (DVFS) techniques dynamically adjust the voltage and frequency of processors based on workload demands, reducing energy consumption during periods of low utilization or less computationally intensive tasks.

- Power-aware resource management means monitoring and managing the power consumption of computational resources, such as GPUs and CPUs, to ensure efficient utilization and minimize energy waste. This can involve techniques like power capping, power-aware job scheduling, and dynamic power management.

- Workload characterization and profiling of generative AI workloads' energy consumption and performance characteristics can inform energy-aware scheduling and resource management decisions, enabling more efficient resource allocation and optimization.

- Heterogeneous resource management involves efficiently managing and scheduling workloads across heterogeneous hardware resources, such as CPUs, GPUs, and specialized accelerators, based on their energy efficiency and performance characteristics for different types of workloads.

By employing energy-aware scheduling and resource management techniques, generative AI systems can optimize energy consumption while maintaining performance requirements, reducing their overall environmental impact and enabling more sustainable operations.

Power-Efficient Software Techniques

In addition to hardware-level optimizations, various software techniques can be employed to improve energy efficiency and reduce power consumption in generative AI systems.

- Clock gating is a power optimization technique that disables the clock signal to unused or idle components of a processor or accelerator, reducing dynamic power consumption and improving energy efficiency.

- Power gating involves cutting off the power supply to unused or idle components, reducing static power consumption and leakage currents. This technique can be applied to various components, such as processor cores, memory blocks, or accelerator units when they are not in use.

- Instruction-level parallelism (ILP) optimization can improve energy efficiency by maximizing the utilization of available computational resources and reducing the number of clock cycles required for executing instructions.

- Memory access optimization involves enhancing memory access patterns and data locality to reduce energy consumption by minimizing unnecessary data movement and cache misses, which can be energy-intensive operations.

- Algorithmic optimizations and approximation techniques that trade off accuracy for energy efficiency can be beneficial in certain applications where strict precision is not required.

- Software pipelining techniques can improve energy efficiency by overlapping the execution of multiple instructions or operations, reducing stalls, and improving resource utilization.

- Compiler optimizations, such as loop unrolling, function inlining, and dead code elimination, can improve energy efficiency by reducing redundant computations and optimizing code execution.

Generative AI systems can achieve significant energy savings and reduce their overall environmental impact by combining these power-efficient software techniques with hardware-level optimizations and energy-aware scheduling strategies.

Optimizing for Heterogeneous Hardware Architectures

Modern computing systems often feature heterogeneous hardware architectures, combining various types of processors, accelerators, and specialized hardware components. Optimizing generative AI software for these heterogeneous architectures is crucial for achieving energy efficiency and maximizing performance.

- Heterogeneous computing: Leveraging heterogeneous computing architectures by offloading computationally intensive tasks to specialized hardware accelerators, such as GPUs, TPUs, or FPGAs, can improve energy efficiency and performance compared to running workloads solely on general-purpose CPUs.

- Workload partitioning and mapping: Partitioning generative AI workloads and mapping them to the most suitable hardware components based on their computational characteristics and energy efficiency profiles can optimize overall system performance and energy consumption.

- Heterogeneous scheduling and load balancing: Employing scheduling and load balancing techniques that distribute workloads across heterogeneous hardware resources based on their energy efficiency, performance characteristics, and resource availability can improve overall system efficiency.

- Heterogeneous memory management: Optimizing memory management and data movement across different memory hierarchies and types (e.g., CPU

113

memory, GPU memory, high-bandwidth memory) can reduce energy consumption and improve performance by minimizing unnecessary data transfers.

- Heterogeneous compilation and code generation: Leveraging compilers and code generation tools that optimize code for heterogeneous hardware architectures, including specialized instructions and accelerator intrinsics, can improve energy efficiency and performance.

- Heterogeneous profiling and optimization: Analyzing the performance and energy consumption characteristics of generative AI workloads on heterogeneous hardware can inform optimization strategies and enable more efficient resource utilization.

By optimizing generative AI software for heterogeneous hardware architectures and leveraging specialized accelerators, significant energy savings and performance improvements can be achieved, enabling more sustainable and efficient deployment of these computationally intensive models.

Leveraging Specialized Hardware Accelerators

Specialized hardware accelerators, such as GPUs, TPUs, and FPGAs, can significantly improve the energy efficiency and performance of generative AI workloads by offloading computationally intensive tasks from general-purpose CPUs.

- Graphics processing units, or GPUs, are widely used for accelerating deep learning and generative AI workloads due to their highly parallel architecture and specialized hardware for matrix operations. Optimizing software for efficient GPU utilization, including techniques like batching, model parallelism, and mixed-precision computations, can improve energy efficiency and performance.

- Tensor processing units, or TPUs, are application-specific integrated circuits (ASICs) designed by Google specifically for accelerating machine learning workloads, including generative AI models. TPUs offer high computational density and energy efficiency, making them well-suited for large-scale model training and inference.

- Field-programmable gate arrays (FPGAs) are reconfigurable hardware devices that can be programmed to implement custom hardware architectures and accelerate specific workloads. FPGAs can be optimized for energy-efficient execution of generative AI models by implementing custom hardware designs and leveraging their reconfigurability.

- Accelerator-specific optimization of software for specific accelerator architectures, such as leveraging specialized instructions, intrinsics, and libraries provided by accelerator vendors, can unlock additional performance and energy efficiency gains.

- Heterogeneous computing with accelerators is a technique that combines specialized accelerators with general-purpose CPUs and other hardware components in a heterogeneous computing environment to enable efficient workload partitioning and mapping, further improving overall system performance and energy efficiency.

- Accelerator virtualization and sharing techniques can improve resource utilization and energy efficiency by enabling multiple workloads or users to share accelerator resources, reducing idle time and maximizing utilization.

By leveraging specialized hardware accelerators and optimizing software for their architectures, generative AI systems can achieve significant performance improvements and energy savings, enabling more sustainable and efficient deployment of these computationally intensive models.

Environmentally Friendly Software Engineering Practices

In addition to optimizing the software itself, adopting efficient software engineering practices throughout the development lifecycle is crucial for achieving sustainable and environmentally responsible generative AI systems. This section explores various practices and strategies for sustainable software development, carbon-aware design and development, measuring and monitoring software carbon footprints, and continuous integration and deployment for sustainable software.

Sustainable Software Development Lifecycle

Integrating sustainability considerations into the software development lifecycle is essential for creating environmentally responsible generative AI systems. Several practices can be adopted.

- Sustainable requirements engineering: Incorporating sustainability goals and environmental impact considerations into the requirements gathering and analysis phase ensures that sustainability is a key driver in software development.

- Sustainable software design: Adopting sustainable software design principles, such as modular and composable architectures, efficient resource utilization, and energy-aware design patterns, to create software systems that are inherently more sustainable and energy-efficient.

- Sustainable coding practices: Promoting coding practices that prioritize energy efficiency, resource optimization, and environmental impact reduction, such as using energy-efficient algorithms, minimizing redundant computations, and optimizing memory usage.

- Sustainable testing and validation: Incorporating sustainability metrics and environmental impact assessments into the testing and validation phases, ensuring that the software meets functional requirements and sustainability goals.

117

- Sustainable deployment and operations: Adopting sustainable deployment and operations practices, such as leveraging energy-efficient infrastructure, optimizing resource utilization, and implementing energy-aware monitoring and management strategies.

Case Studies and Best Practices

Real-world Examples of Software Optimization for Sustainable Generative AI

Several organizations and research groups have implemented software optimization techniques to improve the sustainability and efficiency of their generative AI systems. The following are a few real-world examples.

- OpenAI, a leading AI research company, has focused on optimizing the training and inference of large language models like GPT-3. They have employed techniques such as model parallelism, mixed-precision training, and efficient inference strategies to reduce the computational demands and energy consumption of their models.

- Google AI has been at the forefront of developing energy-efficient AI systems, including generative models like DALL-E and LaMDA. They have leveraged specialized hardware accelerators like TPUs and software optimizations like quantization, pruning, and efficient model architectures to improve the sustainability of their AI systems.

- DeepMind, a subsidiary of Alphabet Inc., has been exploring sustainable AI practices, including optimizing the training and deployment of generative models like AlphaFold. They have implemented techniques such as model compression, efficient training algorithms, and energy-aware scheduling to reduce the environmental impact of their AI systems.

- Hugging Face, a leading provider of open-source natural language processing tools and models has focused on optimizing the inference and deployment of large language models. They have developed libraries and tools for model quantization, pruning, and efficient serving strategies, enabling more sustainable deployment of generative AI models.

- Academic and research institutions have been investigating software optimization techniques for sustainable generative AI. For example, researchers at the University of Cambridge have explored energy-efficient model architectures and training algorithms for large language models, and researchers at MIT have investigated efficient inference strategies and hardware acceleration for generative AI models.

Lessons Learned and Best Practices from Industry and Research

Through the experiences and efforts of industry and research organizations, several best practices and lessons learned have emerged for optimizing software for sustainable generative AI.

- Holistic approach: Adopting a holistic approach that combines multiple optimization techniques across model architecture, training algorithms, inference strategies, and software engineering practices is crucial for achieving significant sustainability gains.

- Hardware-software co-design: Closely collaborating between hardware and software teams to co-design optimized hardware architectures and software frameworks tailored for generative AI workloads can unlock further efficiency improvements.

- Continuous optimization: Treating software optimization as an ongoing process, continuously monitoring and refining optimization strategies as models evolve and new techniques emerge, is essential for maintaining sustainability in the face of rapidly advancing generative AI technologies.

- Benchmarking and measurement: Establishing standardized benchmarks and metrics for measuring the computational efficiency, energy consumption, and environmental impact of generative AI systems is crucial for quantifying the effectiveness of optimization techniques and driving progress toward more sustainable AI.

- Open collaboration and knowledge sharing: Fostering open collaboration and knowledge sharing among industry, academia, and research organizations can accelerate the development and adoption of sustainable software practices for generative AI, leveraging collective expertise and resources.

- Sustainability-driven innovation: Embracing sustainability as a driving force for innovation can lead to the development of novel software architectures, algorithms, and techniques specifically designed for energy-efficient and environmentally responsible generative AI systems.

Conclusion

This chapter has explored various software optimization techniques and strategies, including model architecture optimization, efficient training algorithms, optimized software frameworks and libraries, efficient inference strategies, software optimization for energy efficiency, and green software engineering practices.

By combining these techniques and adopting a holistic approach to software optimization, researchers and developers can unlock significant computational savings, reduce energy consumption, and enable more sustainable deployment of generative AI models. Real-world examples and best practices from industry and research highlight the importance of continuous optimization, hardware-software co-design, benchmarking and measurement, and open collaboration in driving progress toward sustainable AI.

Looking ahead, the field of sustainable software practices for generative AI is poised for significant advancements through several emerging trends and research directions. These include co-designing specialized hardware and software to maximize efficiency, applying automated optimization techniques and machine learning to enhance software performance, and exploring quantum computing's potential in generative AI tasks. AWS's acquisition of Annapurna Labs in 2015 provided just that, a way for AWS to manage both the hardware and software stack allowing for more targeted optimizations for AI workloads. Additionally, the development of sustainable AI software methodologies, efforts toward standardization and benchmarking of sustainability metrics, and increased interdisciplinary collaboration are expected to play crucial roles. These areas of focus promise to drive innovation in creating more environmentally friendly and resource-efficient generative AI systems, addressing the growing need for sustainable computing solutions in an increasingly AI-driven world. By embracing these emerging trends and future research directions, the AI community can continue to push the boundaries of sustainable software practices, enabling the responsible development and deployment of generative AI technologies while minimizing their environmental impact and contributing to a more sustainable future.

The next chapter delves into critical aspects of data management and preprocessing for generative AI systems. It covers three main areas: efficient data storage and retrieval, data cleaning and preprocessing strategies, and data augmentation and synthesis techniques. Chapter 5 also explores strategies for managing large datasets, including distributed storage and parallel access methods. It also addresses the challenges of handling noisy, incomplete, or biased data, providing techniques to clean and preprocess data effectively. Finally, the next chapter discusses various approaches to data augmentation and synthetic data generation, examining their benefits and limitations in the context of Generative AI.

CHAPTER 5

Data Management and Preprocessing for Sustainable Generative AI

Data is the lifeblood of generative AI, serving as the foundation upon which these sophisticated models are built and trained. The quality, quantity, and diversity of data directly influence the performance, capabilities, and outputs of generative AI systems. From text and images to audio and video, vast datasets are required to train models that can generate human-like content across various domains. The critical role of data extends beyond mere training; it shapes the AI's understanding of patterns, context, and nuances, ultimately determining the relevance, accuracy, and creativity of its generated outputs. As generative AI continues to evolve and find applications in diverse fields such as content creation, design, scientific research, and problem-solving, the demand for high-quality, representative data grows exponentially, underscoring its pivotal role in advancing AI technology.

I. K. Dua and P. G. Patel, *Optimizing Generative AI Workloads for Sustainability*,
https://doi.org/10.1007/979-8-8688-0917-0_5

123

Efficient data management is not just a technical necessity but a crucial factor in ensuring the sustainability of generative AI systems. As these models grow in size and complexity, their data requirements and computational demands increase dramatically, leading to significant energy consumption and environmental impact. Sustainable data management practices are essential to mitigate these effects, optimizing storage, processing, and retrieval to reduce the carbon footprint of AI operations. See Figure 5-1.

Figure 5-1. *Data storage innovation*

By implementing efficient data pipelines, leveraging compression techniques, and adopting smart caching strategies, organizations can minimize unnecessary data transfers and computations, thereby conserving energy and resources. Moreover, sustainable data management extends to ethical considerations, ensuring responsible data collection, usage, and retention practices that respect privacy and reduce digital waste. As the AI industry grapples with its growing environmental impact, efficient and sustainable data and storage management emerges as a key strategy in balancing technological advancement with ecological responsibility.

Efficient Data Storage and Retrieval for Sustainable IT

In the era of big data and generative AI, efficient data storage and retrieval have become critical components of sustainable IT practices. As organizations grapple with exponentially growing datasets, the environmental impact of data management cannot be overlooked. This section explores sustainable storage solutions, energy-efficient data centers, distributed storage systems, and techniques for reducing data redundancy and storage footprint.

Sustainable Storage Solutions for Large Datasets

The foundation of sustainable data storage lies in selecting appropriate storage technologies and implementing best practices that minimize energy consumption and environmental impact.

Tiered Storage Architecture

Implementing a tiered storage architecture is a key strategy for optimizing data storage efficiency. This approach involves categorizing data based on access frequency and importance and storing it on appropriate tiers of storage media. Data storage management has evolved to incorporate a tiered approach, optimizing performance and cost-effectiveness. This strategy typically consists of three main tiers: hot, warm, and cold. The hot tier is designed for frequently accessed data, utilizing high-performance, energy-efficient solid-state drives to ensure rapid retrieval and processing. Moving down the hierarchy, the warm tier houses less frequently accessed data on more economical, higher-capacity hard disk drives, striking a balance between accessibility and cost. Finally, the cold tier is reserved for rarely accessed data, employing energy-efficient archival storage solutions that prioritize long-term preservation and minimal operational costs. This tiered structure allows organizations to efficiently manage their data, allocating resources based on access frequency and importance, ultimately improving system performance and reducing overall storage expenses. For example, Amazon S3 Intelligent-Tiering automatically moves data between access tiers based on usage patterns, optimizing cost and energy efficiency. This is especially handy for customers and developers who may not be sure of the access patterns and want to still implement some form of tiering. The following image shows how S3 provides different levels of reliability and availability through its storage classes. Each storage class has a different cost structure and guarantees how quickly you can access your data. See Figure 5-2.

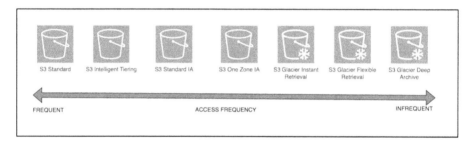

Figure 5-2. *S3 Intelligent-Tiering*

Data Compression and Deduplication

Reducing the overall volume of stored data is a critical component of sustainable storage practices, and two key techniques stand out in this effort: compression and deduplication. Compression utilizes advanced algorithms such as ZSTD or LZ4 to shrink data size without significantly impacting performance, allowing for efficient storage and retrieval. Deduplication, on the other hand, focuses on eliminating redundant data blocks by storing only unique instances, thereby optimizing storage capacity. These methods are not just theoretical concepts but are actively implemented in modern cloud storage solutions. For instance, Azure Data Lake Storage Gen2 exemplifies this approach by supporting transparent compression, which automatically compresses data to reduce storage requirements and associated energy consumption. This practical application demonstrates how these techniques can be seamlessly integrated into storage systems, offering tangible benefits in terms of reduced storage costs, improved energy efficiency, and overall environmental sustainability in data management practices. Figure 5-3 presents a basic compression decompression process.

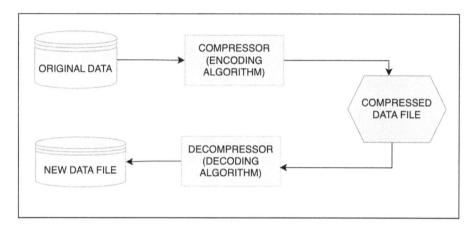

Figure 5-3. *Data compression process*

Software-Defined Storage

Software-defined storage (SDS) solutions abstract storage management from hardware, allowing for more efficient resource utilization and easier implementation of energy-saving policies. For example, Google Cloud's Cloud Storage uses a software-defined approach, enabling features like automatic data migration between storage classes for optimal efficiency.

Energy-Efficient Data Centers and Cloud Storage

Cloud providers have made significant strides in improving data center energy efficiency, offering sustainable options for organizations looking to reduce their environmental impact.

Renewable Energy-Powered Data Centers

Major cloud providers are increasingly powering their data centers with renewable energy sources, demonstrating a significant commitment to environmental sustainability. See Figure 5-4.

Figure 5-4. *Renewable energy*

Amazon Web Services (AWS) has set an ambitious goal to match 100% of the electricity they consume with renewable energy by 2030, which they achieved in 2023 as per their latest sustainability report. Microsoft Azure has been at the forefront of this movement, having achieved 100% renewable energy powering for its operations since 2014, setting an early example for the industry. Google Cloud has maintained carbon neutrality since 2007 and is pushing boundaries further by aiming to run on carbon-free energy around the clock by 2030. These initiatives by leading cloud providers not only reduce their environmental footprint but also drive innovation in clean energy technologies, potentially accelerating the global transition to a more sustainable future in the tech industry and beyond.

Advanced Cooling Technologies

Innovative cooling solutions have emerged as a key strategy to significantly reduce energy consumption in data centers. Free air cooling, which utilizes outside air to cool data centers when ambient temperatures allow, has proven to be an effective method for reducing reliance on traditional air conditioning systems. Liquid cooling takes efficiency a step further by immersing servers in dielectric fluid, enabling more efficient heat dissipation than air-based cooling methods. These technologies are not just theoretical but are being implemented by major cloud providers. For instance, AWS implements evaporative cooling in their data centers, Google Cloud has leveraged advanced machine learning algorithms to optimize its data center cooling systems, resulting in impressive energy savings of up to 40% for cooling operations. This approach demonstrates how cutting-edge technology can be applied to address the critical challenge of energy efficiency in data centers, combining innovative hardware solutions with sophisticated software optimization to achieve substantial reductions in power consumption and environmental impact.

Energy-Efficient Hardware

Cloud providers continually upgrade to more energy-efficient hardware to improve performance and reduce power consumption. ARM-based processors, such as AWS Graviton, offer better performance per watt than traditional x86 processors, making them an increasingly popular choice for cloud infrastructure. See Figure 5-5.

Figure 5-5. *AWS Graviton chip*

Solid-state drives are replacing hard disk drives for frequently accessed data, which reduces power consumption and improves overall performance. These hardware upgrades are crucial in enhancing the energy efficiency of data centers and cloud operations.

Distributed Storage Systems and Their Environmental Impact

Distributed storage systems offer improved scalability and reliability but require careful design to minimize environmental impact.

Geo-Distributed Storage

Geo-distributed storage, which involves storing data across multiple geographic locations, can improve performance and resilience but may increase energy consumption due to data replication and transfer. To mitigate this, best practices include implementing intelligent data placement algorithms to minimize cross-region data transfers and using content delivery networks (CDNs) like Amazon Cloudfront to cache frequently accessed data closer to users.

Edge Storage

Edge storage is another approach that stores and processes data closer to its source, significantly reducing network traffic and associated energy consumption. This method is particularly useful for applications that require low latency or handle large volumes of data at the edge of the network. AWS Outposts extends AWS infrastructure to on-premises locations, enabling local data processing and storage for edge use cases.

Erasure Coding

Erasure coding provides data redundancy with less storage overhead than traditional replication methods. This technique improves data durability while minimizing storage requirements and associated energy consumption, making it an environmentally friendly option for large-scale distributed storage systems. Google Cloud Storage uses erasure coding to provide durability while minimizing storage requirements and associated energy consumption.

Techniques for Reducing Data Redundancy and Storage Footprint

Minimizing unnecessary data duplication and storage is crucial for sustainable IT practices.

Data Lifecycle Management

Data lifecycle management involves implementing automated data retention and deletion policies, ensuring that unnecessary data doesn't consume storage resources indefinitely. This approach helps organizations maintain control over their data growth and storage costs. Amazon S3 Lifecycle configurations allow the automatic transitioning of objects between storage classes and the expiration of objects based on defined rules.

Thin Provisioning

Thin provisioning is another technique that allocates storage space on-demand rather than pre-allocating large volumes, helping to optimize storage utilization. This method is particularly useful in virtualized environments with dynamic and unpredictable storage needs. Azure Managed Disks support thin provisioning, allowing for more efficient use of storage resources.

Data Normalization and Denormalization

Balancing data normalization and denormalization is important for optimizing storage efficiency and application performance. While normalization reduces redundancy, denormalization can improve query performance. Finding the right balance depends on the specific requirements of each application and dataset.

Metadata Management

Efficient metadata management can significantly reduce storage requirements, especially for large-scale unstructured data. Using compact, efficient metadata formats and implementing metadata indexing allows faster searches without scanning entire datasets. This approach is particularly beneficial for organizations dealing with massive amounts of unstructured data, such as media files or scientific datasets. Google Cloud Datastore provides a fully managed, schemaless database for storing non-relational data, with efficient indexing capabilities for optimized queries. AWS Lake Formation is combination with S3 and AWS Glue Data Catalog can be used to effectively manage and maintain metadata.

Data Cleaning and Preprocessing Strategies for Sustainable IT

In sustainable IT and efficient data management, data cleaning and preprocessing play a crucial role in optimizing the performance of AI models while minimizing computational costs and energy consumption. This section explores the importance of clean data, automated cleaning techniques, energy-efficient preprocessing pipelines, and the balance between data quality and computational costs.

Importance of Clean Data for Model Efficiency

Data quality plays a crucial role in the efficiency and sustainability of machine learning operations. Clean data significantly reduces computational overhead by minimizing the need for complex data handling and error correction during model training and inference, leading to lower computational requirements and energy consumption. High-quality data results in more accurate models, reducing the need for retraining and fine-tuning, both computationally intensive processes. Clean data also allows models to converge more quickly during training, reducing the overall energy consumption of the training process. From a storage perspective, eliminating redundant, irrelevant, or erroneous data results in cleaner datasets that require less storage space, leading to reduced energy consumption in data centers. Furthermore, clean

133

data enhances data transfer efficiency by minimizing unnecessary data transfers, thereby reducing network traffic and associated energy costs. These combined benefits of clean data contribute significantly to more sustainable and environmentally friendly machine learning practices.

Automated Data Cleaning Techniques to Reduce Manual Effort

Automating data cleaning processes is essential for sustainable IT practices, as it reduces manual effort, improves consistency, and allows for more efficient use of computational resources. AWS Data Wrangler is a tool within the Amazon Sagemaker ecosystem that helps implement data pre-processing steps mentioned below in a simple drag and drop manner using a UI. Here are some key automated data-cleaning techniques.

Outlier Detection and Handling

Outlier detection and handling are crucial steps in data cleaning that can be automated using various methods. Statistical approaches, such as Z-score or interquartile range (IQR) algorithms, can be employed to automatically identify and handle outliers in datasets. Figure 5-6 graphs an outlier.

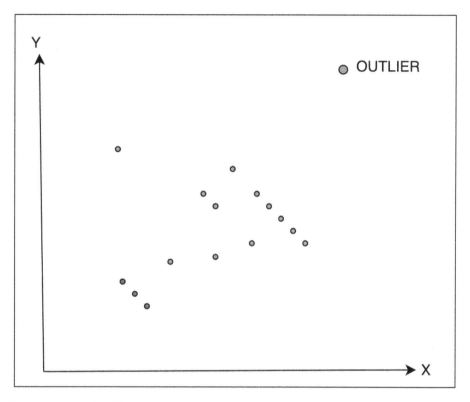

Figure 5-6. *Outlier*

For more complex, high-dimensional data, machine learning-based approaches like clustering algorithms or autoencoders can be utilized to detect anomalies. These automated methods help ensure that outliers are consistently and efficiently identified and addressed, improving the overall quality of the dataset.

Missing Value Imputation

Automated missing value imputation can significantly enhance data completeness and quality. Simple imputation methods can automatically fill missing values with the mean, median, or mode of the respective feature, providing a quick and straightforward solution.

For more sophisticated approaches, advanced imputation techniques using machine learning models such as K-Nearest Neighbors (KNN) or regression can be employed to predict missing values based on other features in the dataset. See Figure 5-7.

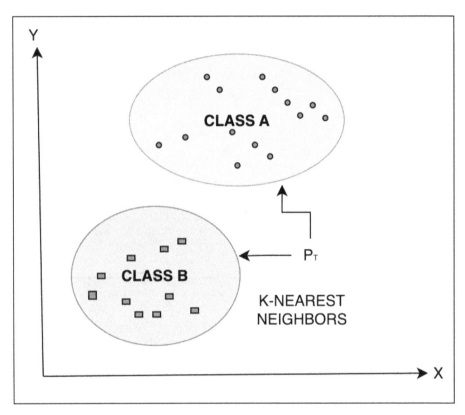

Figure 5-7. *KNN technique*

These automated imputation methods help maintain data integrity and enable more comprehensive analyses.

Duplicate Detection and Removal

Efficient duplicate detection and removal are essential for maintaining data accuracy and reducing redundancy. Hash-based techniques can be implemented to generate hash values for records, allowing for quick identification and removal of exact duplicates. For detecting near-duplicate records, fuzzy matching algorithms like Levenshtein distance or cosine similarity can be employed. These automated approaches ensure consistent and thorough duplicate handling, improving data quality and reducing storage requirements.

Data Type Conversion and Standardization

Automating data type conversion and standardization processes can greatly improve data consistency and usability. Scripts can be developed to automatically detect and convert data types, such as transforming strings to dates or numbers to appropriate numeric types. Additionally, implementing rule-based systems for unit standardization ensures that units of measurement are consistently formatted across the dataset. These automated conversions and standardizations enhance data interoperability and reduce the likelihood of errors in subsequent analyses.

Text Normalization

Automated text normalization is crucial for ensuring consistency in textual data. Developing pipelines for automated text cleaning, including lowercase conversion, punctuation removal, and special character handling, can significantly improve text data quality. Furthermore, integrating spell-check libraries to automatically correct common misspellings enhances the accuracy and uniformity of textual information. These automated text normalization processes contribute to more reliable and consistent textual data, facilitating improved natural language processing and text analysis tasks.

Energy-Efficient Preprocessing Pipelines

Designing energy-efficient preprocessing pipelines is crucial for sustainable IT practices. Here are strategies to optimize preprocessing for reduced energy consumption.

Incremental Processing

Implementing streaming data processing techniques allows for handling data in small batches, reducing memory usage and enabling more efficient resource allocation. This approach can be complemented by using incremental learning algorithms that update models with new data without reprocessing the entire dataset, further optimizing computational resources.

Parallel Processing

Leveraging distributed computing frameworks like Apache Spark or Dask enables the parallelization of data preprocessing tasks across multiple nodes, improving efficiency and reducing overall processing time, as seen in Figure 5-8. Additionally, implementing multithreading for I/O-bound preprocessing tasks maximizes CPU utilization, further enhancing performance.

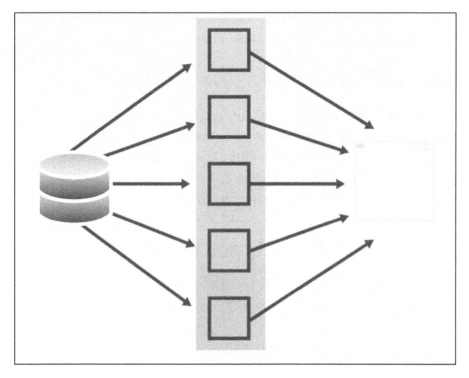

Figure 5-8. *Parallelization techniques*

In-Memory Processing

Utilizing in-memory data processing techniques minimizes disk I/O operations, which are energy-intensive. Implementing caching mechanisms to store frequently accessed intermediate results reduces redundant computations, leading to improved efficiency and reduced energy consumption.

Efficient Data Representations

Compact data formats like Parquet or ORC that offer efficient compression and encoding schemes can significantly reduce storage requirements and processing time. Implementing feature hashing techniques to reduce the

139

dimensionality of categorical variables saves memory and computational resources, contributing to overall efficiency.

Adaptive Preprocessing

Developing smart preprocessing pipelines that can adapt to the characteristics of the input data ensures that only necessary transformations are applied, avoiding unnecessary computations. Implementing early stopping mechanisms in iterative preprocessing algorithms optimizes resource usage by preventing excessive processing.

Balancing Data Quality and Computational Costs

Achieving a balance between data quality and computational costs is essential for sustainable IT practices. Here are strategies to optimize this trade-off.

Selective Cleaning

Selective cleaning involves using intelligent sampling techniques to clean a representative subset of the data and then applying learned cleaning rules to the entire dataset. Prioritizing cleaning efforts on features that have the most impact on model performance, using techniques like feature importance analysis, ensures efficient use of resources.

Progressive Data Quality Improvement

Progressive data quality improvement implements iterative cleaning processes that gradually enhance data quality while monitoring the impact on model performance. Active learning techniques can be used to identify and prioritize the most valuable instances for manual review and cleaning, optimizing human intervention.

Approximate Computing

Approximate computing techniques can be employed in preprocessing steps where slight inaccuracies are tolerable, reducing computational costs. Probabilistic data structures like Bloom filters can be used for efficient duplicate detection in large datasets, balancing accuracy with resource usage.

Transfer Learning for Data Cleaning

Transfer learning for data cleaning leverages pre-trained models or transfer learning techniques to reduce the computational cost of developing data cleaning models for new datasets. This approach allows for more efficient adaptation to new data-cleaning tasks.

Quality-Aware Resource Allocation

Quality-aware resource allocation involves developing systems that dynamically allocate computational resources based on the current quality of the data and the desired quality threshold. Implementing early termination strategies for preprocessing tasks when diminishing returns in quality improvement are observed helps optimize resource utilization.

Continuous Monitoring and Optimization

Continuous monitoring and optimization are crucial for maintaining an optimal balance. Implementing monitoring systems to track the impact of data cleaning on both model performance and computational costs allows for ongoing refinement. Automated optimization techniques can continuously adjust the balance between cleaning efforts and resource utilization, ensuring long-term sustainability in data preprocessing practices.

By implementing these data cleaning and preprocessing strategies, organizations can significantly improve the efficiency of their AI models while minimizing the environmental impact of their IT operations. The key lies in automating processes, optimizing resource usage, and continuously balancing data quality with computational costs, as seen in Figure 5-9.

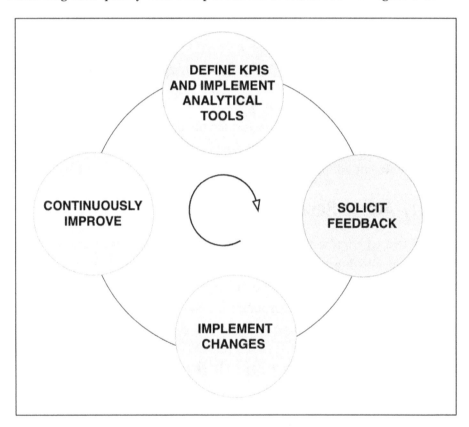

Figure 5-9. *Continuous improvement*

Data Augmentation and Synthesis

Data augmentation and synthesis play crucial roles in sustainable AI development by reducing the need for extensive real-world data collection and optimizing model training processes. This section explores sustainable approaches to data augmentation, synthetic data generation techniques, and the trade-offs between augmentation and model training costs.

Geometric Transformations

Geometric transformations offer computationally efficient methods to significantly increase dataset diversity without requiring additional real-world data collection. These techniques include rotation, which creates new perspectives by rotating images or 3D data points; flipping, which doubles the dataset size by horizontally or vertically flipping images; scaling, which adjusts the size of objects within images or scales numerical data; and translation, which shifts data points or image contents to create new samples. By applying these transformations, datasets can be expanded and diversified, enhancing model robustness and generalization capabilities.

Color Space Transformations

Color space transformations help models become more robust to varying lighting and color conditions, reducing the need for extensive real-world data collection under different environments. These techniques include brightness adjustment to simulate different lighting conditions, contrast modification to create variations, color jittering to randomly adjust images' hue, saturation, and value, and grayscale conversion to add further variety. By applying these transformations, models can learn to handle a wider range of visual inputs, improving their performance across diverse real-world scenarios.

Noise Injection

Noise injection techniques help models become more resilient to real-world data imperfections and sensor noise, improving generalization without requiring additional data collection. Methods include adding

Gaussian noise to images or numerical data, introducing salt-and-pepper noise with random black-and-white pixels in images, and applying speckle noise by multiplying data points with random noise values. These techniques simulate real-world imperfections, enabling models to learn more robust features and perform better on noisy or imperfect data.

Mixing and Patching

Mixing and patching techniques create diverse and challenging training samples, potentially reducing the need for larger datasets and extensive data collection efforts. These methods include Mixup, which creates new samples by linearly combining existing samples and their labels; CutMix, which replaces rectangular regions of images with patches from other images; and MosaicMix, which combines multiple images into a single mosaic-like image. These approaches generate novel training examples to help models learn more generalizable features and improve overall performance.

Time Series and Sequential Data Augmentation

Time series and sequential data augmentation methods are particularly useful for augmenting sensor data, financial time series, or textual data, reducing the need for extensive real-world data collection in these domains. Techniques include time warping to stretch or compress time series data along the time axis, magnitude warping to adjust the magnitude of time series data points, permutation to shuffle segments of sequential data, creating new sequences, and jittering to add small random noise to time series data points. These methods help models learn temporal patterns more effectively and become more robust to variations in sequential data.

Natural Language Processing (NLP) Augmentation

NLP augmentation techniques can significantly increase the diversity of textual datasets without requiring additional data collection or annotation efforts. Methods include synonym replacement to substitute words with their synonyms, back-translation to translate text to another language and back to the original, text generation using pre-trained language models to create new text samples, and word swapping to randomly swap words within a sentence. These techniques help create diverse linguistic variations, enabling models to better understand and process natural language across different contexts and styles.

Audio Augmentation

Audio augmentation techniques enhance the diversity and robustness of audio datasets without necessitating extensive new recordings. These methods include time stretching to slow down or speed up audio without changing the pitch, pitch shifting to adjust the pitch of audio samples up or down, adding background noise to mix clean audio with various environmental sounds, and room simulation to apply reverberation effects that mimic different acoustic environments. By applying these transformations, audio datasets can be expanded to cover a wider range of acoustic conditions, improving model performance in various real-world audio processing tasks.

These techniques can create diverse audio datasets for speech recognition, music classification, or environmental sound detection tasks, reducing the need for extensive real-world audio recordings.

Synthetic Data Generation to Reduce Real-World Data Collection

Generative Adversarial Networks

Generative adversarial networks (GANs) have emerged as a powerful tool for synthetic data generation across various domains. In image generation, GANs can create realistic images of faces, objects, or scenes, significantly expanding visual datasets. For natural language processing tasks, GANs can generate synthetic text data, enhancing the diversity of linguistic samples. In time series analysis, GANs are utilized to create synthetic data for financial or sensor data analysis. The high-quality synthetic data produced by GANs closely resembles real-world data, substantially reducing the need for extensive data collection and annotation efforts. This capability makes GANs an invaluable asset in data-hungry machine learning applications. Figure 5-10 represents the structure of GANs.

Figure 5-10. *Structure of GANs*

Variational Autoencoders

Variational autoencoders (VAEs) offer a probabilistic approach to data generation, allowing for controlled synthesis of diverse samples. In image synthesis, VAEs can generate new images by sampling from the learned latent space, providing a flexible method for expanding image datasets. VAEs are employed to create coherent text passages for text generation, enhancing linguistic diversity in natural language processing tasks. Additionally, VAEs are useful in anomaly detection, where they can generate synthetic anomalies for imbalanced datasets, improving model performance in detecting rare events. VAEs are used in the popular StabilityAI Stable Diffusion XL base model for image construction. The probabilistic nature of VAEs enables fine-grained control over the generated samples, making them particularly valuable in scenarios requiring specific data characteristics. Figure 5-11 illustrates variational encoders.

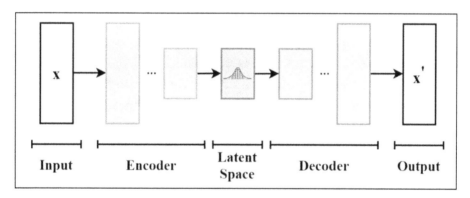

Figure 5-11. *Variational encoders*

Physics-Based Simulations

Physics-based simulations provide a powerful means of generating synthetic data, especially in domains where real-world data collection is expensive, dangerous, or ethically challenging. In robotics, these simulations can model robot movements and interactions in virtual environments, facilitating the development and testing of robotic systems. For autonomous vehicles, physics-based simulations generate synthetic driving scenarios and sensor data, enabling extensive testing without real-world risks. In medical imaging, these techniques can create synthetic images for rare conditions or augment limited datasets, addressing the scarcity of data in certain medical fields. The high degree of realism achievable through physics-based simulations makes them an invaluable tool in data-sensitive and high-stakes domains.

Procedural Generation

Procedural generation techniques offer a powerful approach to creating vast amounts of diverse data with controllable parameters, significantly reducing the need for manual creation or collection of assets. In 3D modeling, these methods can generate synthetic objects and environments for computer vision tasks, providing a rich variety of visual data. For texture and material creation, procedural generation produces diverse textures and material properties for rendering, enhancing the realism of synthetic visual data. In terrain generation, these techniques can produce synthetic landscapes for geospatial analysis or game development, offering a wide range of environmental data. The ability to rapidly generate large volumes of diverse, parameterized data makes procedural generation particularly valuable in scenarios requiring extensive and varied datasets.

Rule-Based Systems

Rule-based systems are efficient for producing large volumes of structured data, particularly useful for specific domains with well-defined constraints. In synthetic text generation, these systems can create structured text data based on predefined templates and rules, enabling the rapid production of domain-specific textual content. For data augmentation, rule-based approaches apply domain-specific transformations to existing data, creating new samples that adhere to known constraints. Rule-based systems can generate synthetic test cases in software testing and quality assurance, ensuring comprehensive coverage of potential scenarios. Figure 5-12 illustrates a rule-based system and what the process could look like.

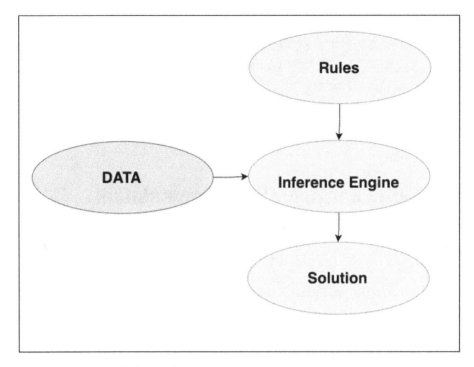

Figure 5-12. *Rule-based system*

The deterministic nature of rule-based systems makes them particularly suitable for generating data in domains where strict adherence to specific patterns or rules is crucial.

Hybrid Approaches

Hybrid approaches combine multiple synthetic data generation techniques to leverage the strengths of each method. GAN–simulation fusion integrates GAN-generated images with physics-based simulations, enhancing the realism of synthetic data while maintaining physical accuracy. VAE–rule-based systems use VAEs to generate base samples, which are refined using rule-based transformations, allowing diversity and adherence to specific constraints. Procedural–GAN pipelines employ procedural generation for base content creation, followed by GAN refinement for added realism. These hybrid methods offer a powerful and flexible approach to synthetic data generation, enabling the creation of highly realistic and diverse datasets that can be tailored to specific application requirements.

Hybrid approaches leverage the strengths of multiple techniques to produce high-quality, diverse synthetic data that closely mimics real-world complexity.

Trade-offs between Data Augmentation and Model Training Costs

Computational Overhead

Data augmentation presents a trade-off in computational resources. On the positive side, it can reduce the need for larger datasets, potentially decreasing storage and data management costs. However, complex augmentation techniques may increase preprocessing time and computational requirements during training. To mitigate these challenges, it's advisable to implement efficient, parallelized augmentation pipelines and consider on-the-fly augmentation during training to balance computational costs. This approach allows for a more dynamic and resource-efficient augmentation process.

Model Complexity

Augmented datasets can have varying impacts on model complexity. They can improve model generalization, potentially allowing for simpler model architectures. However, excessive augmentation may require more complex models to capture the increased data variability. To address this, it's crucial to carefully tune the augmentation strategy to match the model's capacity and the task's requirements. This balanced approach ensures that the benefits of augmentation are realized without unnecessarily complicating the model architecture.

Training Time

The impact of augmentation on training time is multifaceted. It can lead to faster convergence by exposing the model to more diverse examples, potentially reducing the number of epochs required for optimal performance. However, the increased dataset size due to augmentation may prolong overall training time. To mitigate this, implementing adaptive augmentation strategies that focus on the most effective transformations as training progresses can be beneficial. This approach optimizes the augmentation process, ensuring that the most valuable transformations are prioritized throughout the training cycle.

Energy Consumption

Data augmentation can have both positive and negative effects on energy consumption. The reduced need for real-world data collection can lower energy costs associated with data acquisition and storage. However, intensive augmentation processes may increase energy consumption during training. To address this, it's important to optimize augmentation pipelines for energy efficiency and consider using specialized hardware accelerators for augmentation tasks. This approach helps balance the energy savings from reduced data collection with the potential increase in computational energy requirements.

Data Quality and Bias

Augmentation can be a powerful tool for addressing dataset imbalances and improving the representation of underrepresented classes. However, poorly designed augmentation strategies may introduce biases or unrealistic samples. To mitigate these risks, it's essential to regularly evaluate the quality and diversity of augmented data and validate model performance on non-augmented test sets. This ongoing assessment ensures that the augmentation process enhances data quality without introducing unintended biases or artifacts.

Model Robustness

Diverse augmented datasets can significantly improve model robustness to various real-world conditions. However, over-reliance on synthetic data may lead to poor generalization to real-world scenarios. To strike the right balance, it's important to combine augmented data with real-world samples and validate models on diverse, non-augmented test sets. This approach ensures that the model benefits from the diversity introduced by augmentation while maintaining its ability to perform well on real-world data.

Development and Maintenance Costs

Effective augmentation strategies can reduce the need for ongoing data collection and annotation efforts, potentially lowering long-term costs. However, developing and maintaining sophisticated augmentation pipelines may require additional expertise and resources. Investing in reusable, modular augmentation frameworks that can be easily adapted to different projects and domains is recommended to address this challenge. This approach allows organizations to leverage their augmentation capabilities across multiple projects, maximizing the return on investment in augmentation technologies.

By carefully considering these trade-offs and implementing appropriate mitigation strategies, organizations can leverage data augmentation and synthesis techniques to build more sustainable AI systems while optimizing computational resources and energy consumption.

Optimizing Data Pipelines

Efficient data pipelines are crucial for sustainable AI development, as they directly impact energy consumption, resource utilization, and overall system performance. This section explores strategies for designing energy-efficient data pipelines, techniques for reducing data transfer and network usage, and caching strategies to minimize redundant computations.

Modular Pipeline Architecture

Implement a modular design allowing easy updates and optimizations of individual components. This approach facilitates the use of a microservices architecture, enabling independent scaling of pipeline stages based on workload demands. To ensure consistent environments and efficient resource allocation, leverage containerization technologies such as Docker.

Parallel Processing and Resource Allocation

Utilize distributed computing frameworks like Apache Spark or Dask for parallel data processing while implementing multithreading and multiprocessing techniques to maximize CPU utilization. Consider using GPU acceleration for computationally intensive tasks, such as image processing or deep learning operations. Implement autoscaling mechanisms to adjust computational resources based on workload and use serverless computing platforms for intermittent or bursty workloads to minimize idle resource consumption. Employ job scheduling algorithms that optimize resource utilization across multiple pipeline stages.

Data-Aware Processing and Hardware Selection

Implement early filtering and data reduction techniques to minimize unnecessary processing of irrelevant data. Use adaptive sampling techniques to process representative subsets of large datasets when appropriate and leverage incremental processing methods to handle streaming data efficiently. When selecting hardware, choose energy-efficient CPUs and GPUs optimized for data processing workloads. Consider utilizing specialized hardware accelerators, such as FPGAs or TPUs, for specific pipeline stages, and explore ARM-based processors for edge computing nodes in distributed pipelines. For example, AWS Graviton Series which are ARM based processors are 60% more energy efficient as compared to comparable X86 processors.

Pipeline Monitoring and Optimization

Implement comprehensive monitoring to identify bottlenecks and inefficiencies in the pipeline. Utilize automated performance tuning tools to optimize pipeline configurations and regularly review and refactor pipeline components to incorporate more efficient algorithms and technologies. This ongoing monitoring and optimization process ensures that your pipeline remains efficient and effective as workloads and technologies evolve.

Techniques for Reducing Data Transfer and Network Usage

Data Compression and Serialization

Implementing lossless compression algorithms like Snappy or LZ4 for general-purpose data compression is crucial for optimizing data transfer. Employing domain-specific compression techniques, such as quantization for neural network weights or specialized image compression formats, can significantly reduce data size. Adaptive compression strategies that balance compression ratio with computational overhead should also be considered. For data serialization, efficient formats like Protocol Buffers or Apache Avro can minimize data size during transfer. Implementing schema evolution techniques maintains backward compatibility while optimizing data structures. Using binary serialization rather than text-based formats for numeric and structured data further enhances efficiency.

Incremental Processing and Data Locality

Incremental data processing techniques, such as change data capture, allow for processing only modified or new data, reducing overall data transfer. Delta encoding can only transfer differences between data versions, while incremental learning algorithms update models without reprocessing entire datasets. Optimizing data locality by co-locating data processing tasks with data storage minimizes data movement. Implementing data partitioning strategies that align with processing patterns reduces cross-node data transfers. Judicious use of data replication balances availability with storage and synchronization costs.

Efficient Data Formats and Network Protocols

Utilizing columnar storage formats like Apache Parquet for analytical workloads and implementing data denormalization techniques can reduce join operations and data fetches. Memory-mapped file formats enable fast random access to large datasets. For network protocol optimization, leveraging HTTP/2 or HTTP/3 reduces latency and improves multiplexing in web-based data transfers. Implementing WebSocket or gRPC allows efficient bidirectional streaming communication, while UDP-based protocols can be used for real-time data streaming where appropriate.

Edge Computing and Analytics

Edge computing plays a crucial role in optimizing data transfer by processing and filtering data at the edge, reducing the volume of data transferred to central servers. Implementing federated learning techniques enables model training across distributed edge devices. Edge analytics generate insights locally, transmitting only aggregated results, which significantly reduces data transfer requirements while maintaining the value of the insights generated from the data.

Caching Strategies to Minimize Redundant Computations

Caching is a critical strategy one can use to minimize redundant computations. Figure 5-13 highlights the process entails.

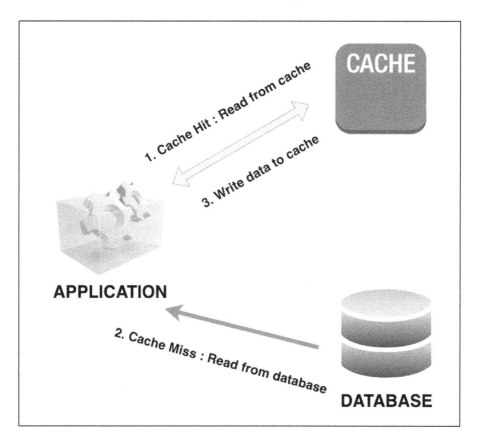

Figure 5-13. *Caching process*

Multilevel and Intelligent Caching

Implementing multilevel caching is crucial for optimizing data access. In-memory caching using technologies like Redis or Memcached provides fast access to frequently used data. Distributed caching systems can share cached data across multiple nodes in a cluster. In contrast, disk-based caching is useful for larger datasets that don't fit in-memory but benefit from faster-than-network access. To ensure data freshness, intelligent cache invalidation strategies are essential. These include time-based expiration for cached items, version-based cache invalidation to update

157

cached data only when source data changes and event-driven cache invalidation triggered by specific system events or data updates.

Predictive and Computation Caching

Predictive caching leverages machine learning models to anticipate and pre-cache data likely to be needed soon. Historical access patterns can be used to optimize cache replacement policies, while context-aware caching considers user behavior or application state. Computation caching focuses on storing intermediate results of complex computations to avoid redundant processing. This includes implementing memorization techniques for recursive or repetitive function calls and using probabilistic data structures like Bloom filters to quickly check for cache membership.

Partial Result and Adaptive Caching

Caching partial query results can significantly speed up similar future queries. Implementing materialized views for frequently accessed data subsets and using incremental view maintenance techniques efficiently updates cached partial results. Adaptive caching policies are crucial for optimizing cache performance. These include dynamic cache sizing based on workload characteristics and available resources, using adaptive replacement algorithms like Adaptive Replacement Cache (ARC) to balance recency and frequency of access, and employing machine learning techniques to dynamically adjust caching policies based on observed access patterns.

Collaborative Caching

In distributed systems, collaborative caching can greatly enhance performance. Implementing peer-to-peer caching mechanisms allows nodes to share cached data. Content-based networking techniques can route requests to the nearest cache containing the required data.

Collaborative filtering algorithms can also predict and pre-cache relevant data across user groups, further optimizing data access and reducing latency in distributed environments.

Organizations can significantly reduce energy consumption, minimize network usage, and improve overall system efficiency by implementing these optimization strategies for data pipelines. These approaches not only contribute to more sustainable AI development but also enhance system performance and scalability.

Data Compression and Dimensionality Reduction

Data compression and dimensionality reduction techniques are crucial for sustainable AI development, as they help minimize storage requirements, reduce computational complexity, and optimize data transfer. This section explores sustainable data compression techniques, dimensionality reduction methods for improved efficiency, and strategies for balancing compression and model performance.

Sustainable Data Compression Techniques

Lossless Compression Algorithms and Lossy Compression for Multimedia

Lossless compression algorithms are essential for preserving data integrity while achieving significant compression ratios. Huffman Coding assigns shorter codes to more frequent symbols, reducing overall data size. The Lempel-Ziv-Welch (LZW) algorithm builds a dictionary of repeated patterns for efficient encoding, while the Burrows-Wheeler Transform (BWT) rearranges data to group similar characters, improving compressibility. These algorithms are suitable for various data types where data integrity is crucial.

For multimedia data, lossy compression techniques can dramatically reduce storage and bandwidth requirements while maintaining acceptable quality for most applications. JPEG for images uses discrete cosine transform (DCT) and quantization to compress images with minimal visual quality loss. H.264/H.265 for video employs inter-frame prediction and motion compensation to achieve high compression ratios. MP3/AAC for audio utilizes psychoacoustic models to remove imperceptible audio information, reducing file sizes.

Domain-Specific Compression and Neural Network Model Compression

Domain-specific compression techniques leverage the unique characteristics of particular data types to achieve higher compression ratios than general-purpose algorithms. For genomic data, reference-based compression or specialized algorithms like CRAM are used to efficiently store genomic sequences. Time series compression employs techniques like Piecewise Aggregate Approximation (PAA) or Symbolic Aggregate Approximation (SAX) for compact representation. Point cloud compression utilizes octree-based methods or deep learning-based approaches for efficient 3D point cloud representation.

Neural network model compression techniques can significantly reduce the size and computational requirements of AI models, enabling deployment on resource-constrained devices and reducing energy consumption. Quantization reduces the precision of model weights and activations, pruning removes unnecessary connections or neurons without significant performance loss, and knowledge distillation trains smaller, more efficient models to mimic the behavior of larger, more complex models.

Dimensionality Reduction Methods for Improved Efficiency

Various dimensionality reduction methods can improve efficiency in AI systems. Principal component analysis (PCA) is a linear technique that identifies the most important features in high-dimensional data. t-distributed stochastic neighbor embedding (t-SNE) is a non-linear technique that preserves local structure in high-dimensional data, which is particularly useful for visualization. Autoencoders are neural network-based approaches that learn compact representations of input data. Random projection is a computationally efficient method that projects high-dimensional data onto a lower-dimensional subspace. Locally linear embedding (LLE) preserves local relationships between data points, while uniform manifold approximation and projection (UMAP) combines ideas from manifold learning and topological data analysis for efficient dimensionality reduction.

Balancing Compression and Model Performance

To optimize AI systems, it's crucial to balance compression and model performance. Adaptive compression techniques adjust based on data characteristics and model requirements, using different compression levels for different parts of the data or model. Compression-aware training incorporates compression techniques during the model training process. Continuous performance monitoring and automated testing pipelines ensure compressed models meet accuracy thresholds. Hybrid approaches combine multiple compression and dimensionality reduction techniques for optimal balance. Task-specific optimization tailors strategies to specific AI tasks and domains, considering trade-offs between model size, inference speed, and accuracy. Incremental compression techniques progressively reduce model size or data dimensionality, allowing for fine-tuning at each stage to maintain performance.

161

By carefully applying these data compression and dimensionality reduction techniques, organizations can significantly reduce storage requirements, computational complexity, and energy consumption in AI systems while maintaining acceptable levels of performance.

Ethical Considerations in Data Management

As organizations strive for sustainable AI development, ethical considerations in data management become increasingly important. This section explores responsible data collection and usage practices, privacy-preserving techniques and their sustainability impact, and strategies for balancing data retention with environmental concerns.

Responsible Data Collection and Usage

Data Minimization and Informed Consent

Data minimization is crucial for reducing storage requirements and potential privacy risks. Organizations should collect only the data necessary for the intended purpose and implement data auditing processes to identify and remove unnecessary or redundant data. Alongside this, informed consent practices are essential. Clearly communicating data collection purposes and usage to individuals and implementing robust consent management systems that allow users to control their data preferences ensures transparency and user empowerment.

Data Quality, Accuracy, and Fairness

Maintaining high-quality, accurate datasets is vital for responsible AI development. Implementing data validation and cleaning processes, along with regular updates and verification, ensures data relevance and accuracy. Fairness and bias mitigation are equally important. Assessing datasets for potential biases and implementing techniques to mitigate unfair representations, along with using diverse and representative data sources, helps ensure AI models are trained on inclusive datasets.

Transparency, Explainability, and Ethical Data Sharing

Transparency in AI systems is achieved by documenting data collection methodologies, processing pipelines, and usage policies. Implementing explainable AI techniques provides insights into how data influences model decisions. For ethical data sharing, organizations should develop clear agreements that outline usage restrictions and privacy protections and implement secure data-sharing mechanisms that protect sensitive information during transfers.

Privacy-Preserving Techniques and Their Sustainability Impact

Various privacy-preserving techniques can be employed, each with different sustainability implications. Differential privacy adds controlled noise to datasets or model outputs, balancing privacy guarantees with computational costs. Federated learning trains AI models across decentralized devices or servers, potentially reducing data transfer and centralized storage requirements. Homomorphic encryption allows computations on encrypted data, though its complex schemes may increase energy consumption. Secure multiparty computation enables private joint computations but may incur increased computational and communication costs. Data anonymization and pseudonymization techniques protect individual privacy in released datasets. Zero-knowledge proofs verify truths without revealing additional information, though their computational complexity should be considered.

Balancing Data Retention and Environmental Concerns

Effective data lifecycle management is crucial for balancing retention needs with environmental concerns. Implementing automated retention policies and tiered storage systems can optimize data management. Regular archiving or deletion of obsolete data reduces storage requirements and energy consumption. Applying efficient compression algorithms and deduplication techniques further minimizes storage needs. Edge computing for data preprocessing can reduce data transfer and storage in centralized systems. Sustainable storage technologies and carbon-aware data management practices can significantly reduce environmental impact. Organizations must also balance regulatory compliance with environmental responsibility, advocating for updates to data protection regulations that consider the environmental impacts of data retention.

Monitoring and Measuring Data-Related Energy Consumption

- Effective monitoring of data-related energy consumption is crucial for sustainable data management. Implementing real-time power usage effectiveness (PUE) monitoring systems helps track the ratio of total facility energy to IT equipment energy. Smart PDUs (power distribution units) can measure power consumption at the rack or individual server level. Data center infrastructure management (DCIM) software comprehensively monitors data center power consumption, cooling efficiency, and asset utilization. DCIM tools can create energy consumption heat maps and identify areas for optimization.

- Server-level power monitoring leverages the built-in capabilities of modern servers, such as Intel's Data Center Manager or AMD's Power Cap Manager. Agent-based monitoring solutions collect detailed power consumption data from individual servers. For network power monitoring, network management protocols like Simple Network Management Protocol (SNMP) can collect data from network devices, while specialized tools track energy usage across the entire network infrastructure.

- Storage power monitoring utilizes storage array management interfaces to collect power consumption data from storage systems. Storage resource management (SRM) tools with power monitoring capabilities provide additional insights. Application performance management (APM) tools can correlate application performance with energy consumption, helping identify energy-intensive applications and optimize their resource usage.

- Energy management information systems (EMIS) aggregate energy data from multiple sources, providing comprehensive energy usage analytics. EMIS solutions help set energy performance baselines and track progress toward sustainability goals, offering a holistic view of energy consumption across the data infrastructure.

Key Performance Indicators for Sustainable Data Practices

- Several key performance indicators (KPIs) should be monitored to assess and improve sustainable data practices. Energy efficiency metrics include PUE, which measures the ratio of total facility energy to IT equipment energy; data center infrastructure efficiency (DCiE), calculating the percentage of power used by IT equipment relative to total facility power; and energy reuse effectiveness (ERE), assessing the amount of energy reused outside the data center.

- Carbon footprint metrics are crucial for environmental impact assessment. Carbon usage effectiveness (CUE) measures the total CO_2 emissions caused by total data center energy consumption, while renewable energy factor (REF) calculates the percentage of energy consumed from renewable sources.

- Resource utilization metrics provide insights into operational efficiency. These include server utilization rates (monitoring average CPU, memory, and storage utilization), storage efficiency (tracking data reduction ratios and capacity utilization), and network utilization (measuring bandwidth usage and device utilization rates).

- Thermal management is another critical area for sustainability. Cooling efficiency metrics, such as computer room air conditioning (CRAC) efficiency and temperature distribution, along with airflow management metrics related to hot/cold aisle separation and air recirculation, help optimize energy use for cooling.

- Water usage in data centers is measured through water usage effectiveness (WUE), which tracks the amount of water used for cooling and other operations. E-waste management metrics, including equipment lifecycle tracking and recycling rates, help assess the environmental impact of hardware management practices.

- By implementing comprehensive monitoring systems and tracking these KPIs, organizations can gain a detailed understanding of their data-related energy consumption and environmental impact. This information is crucial for identifying areas for improvement, setting sustainability goals, and measuring progress toward more environmentally responsible data management practices.

Strategies for Continuous Improvement in Sustainable Data Management

- Energy benchmarking is a crucial strategy for continuous improvement in sustainable data management. Organizations should regularly compare their energy performance against industry standards and best practices and participate in energy efficiency certification programs like ENERGY STAR for data centers. This benchmarking process provides valuable insights into areas for improvement and helps set realistic goals for energy efficiency.

- Predictive analytics can play a significant role in optimizing energy consumption. Organizations can proactively manage their energy use by implementing machine learning models to predict energy consumption patterns and identify optimization opportunities. Predictive maintenance techniques can also optimize equipment performance and reduce energy waste, ensuring that data center infrastructure operates at peak efficiency.

- Automated energy management systems powered by AI can automatically adjust data center operations for optimal efficiency. These systems can implement dynamic power capping and workload balancing based on real-time energy consumption data, leading to significant energy savings. Continuous commissioning is another important practice involving regular assessment and fine-tuning of data center systems to ensure they operate efficiently. Automated commissioning tools can continuously monitor and optimize system performance, maintaining high energy efficiency levels over time.

- Integrating green software development practices into the organization's IT processes is crucial for comprehensive energy management. This involves incorporating energy efficiency considerations into the software development lifecycle and implementing tools to measure and optimize the energy consumption of software applications. By addressing energy efficiency at the software level, organizations can reduce the overall energy footprint of their IT operations.

- Employee engagement and training are essential
  for successfully implementing sustainable data
  management practices. Developing training programs
  to educate IT staff on sustainable practices and
  implementing gamification strategies to encourage
  energy-saving behaviors can create a culture of
  sustainability within the organization. Collaboration
  with suppliers and vendors is equally important.
  Working with hardware and software vendors
  to develop more energy-efficient solutions and
  implementing sustainability criteria in procurement
  processes for IT equipment and services can drive
  industry-wide improvements in energy efficiency.

Case Studies: Real-world Examples of Sustainable Data Management in Generative AI

- Leading technology companies have successfully
  implemented sustainable data management practices
  in generative AI. OpenAI tackled the challenge of
  reducing the environmental impact of training large
  language models with GPT-3. By implementing a more
  efficient model architecture and training process, studies
  indicated that they reduced the energy consumption
  of GPT-3 training by 30% compared to its predecessor,
  achieving state-of-the-art performance while
  significantly reducing the carbon footprint of model
  training.

- Google's DeepMind AI for Data Center Cooling
  addressed the challenge of optimizing energy
  consumption in data centers hosting AI workloads. By

deploying a reinforcement learning system to control data center cooling systems autonomously, they reduced cooling energy consumption by 40%, leading to a 15% overall reduction in PUE.

- Microsoft's Project Natick took an innovative approach to addressing the energy-intensive cooling requirements of data centers hosting AI workloads. By deploying underwater data centers powered by offshore renewable energy sources, they achieved a PUE of 1.07, significantly lower than traditional data centers, while leveraging clean energy for AI computations.

- NVIDIA's Selene Supercomputer project focused on building a high-performance computing system for AI research with minimal environmental impact. By designing a highly efficient supercomputer using liquid cooling and optimized power delivery, they achieved top rankings in performance and energy efficiency benchmarks, demonstrating that sustainable practices can be successfully implemented in high-performance AI computing.

These case studies demonstrate that with innovative approaches and a commitment to sustainability, organizations can significantly reduce the environmental impact of their AI and data management operations while maintaining or even improving performance.

Conclusion

Sustainable data management practices are crucial for the responsible development and deployment of generative AI systems. As the field continues to advance, organizations must prioritize efficient data storage, preprocessing, and augmentation techniques to minimize environmental impact while maximizing model performance. Key strategies for sustainable AI data management include:

1. Implementing tiered storage architectures and software-defined storage solutions to optimize resource utilization.

2. Leveraging energy-efficient data centers powered by renewable energy sources and advanced cooling technologies.

3. Utilizing distributed storage systems and edge computing to reduce data transfer and processing energy costs.

4. Employing automated data cleaning techniques and energy-efficient preprocessing pipelines to improve data quality while minimizing computational overhead.

5. Balancing data augmentation and synthesis with model training costs to expand datasets efficiently.

By adopting these practices, organizations can significantly reduce the carbon footprint of their AI operations while continuing to push the boundaries of generative AI capabilities. As the industry evolves, ongoing research and innovation in sustainable data management will be essential to ensure that AI development aligns with broader environmental goals and responsible technology practices.

CHAPTER 6

Model Training and Inference Optimization

Model training and inference optimization is crucial for environmental sustainability in the rapidly expanding field of generative AI. As models grow in complexity and scale, their environmental impact becomes increasingly significant, with training and inference processes consuming substantial computational resources and energy. This chapter dives into a few ways to manage and optimize your models for deployment and inference. Specifically, the focus is on neural networks—a difficult architecture to manage due to its abundant parameters and memory required, as illustrated in Figure 6-1.

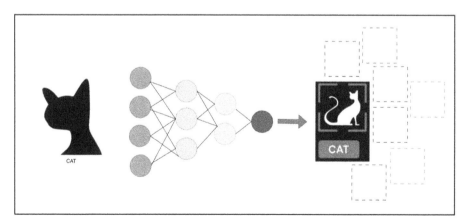

Figure 6-1. *Neural networks*

© Ishneet Kaur Dua and Parth Girish Patel 2024
I. K. Dua and P. G. Patel, *Optimizing Generative AI Workloads for Sustainability*,
https://doi.org/10.1007/979-8-8688-0917-0_6

Optimizing these processes is essential to minimize the carbon footprint of AI workloads while maintaining high performance. By implementing efficient training strategies, fine-tuning hyperparameters, and employing advanced inference techniques, organizations can significantly reduce their AI models' energy consumption and resource utilization. This contributes to environmental conservation and aligns with global sustainability goals, ensuring that the benefits of AI can be realized without compromising our planet's future. Cloud providers like Amazon Web Services (AWS), Microsoft Azure, and Google Cloud Platform (GCP) offer various tools and services to facilitate these processes, ensuring that models can be trained and deployed effectively while minimizing computational costs and energy consumption.

Distributed and Parallel Training Strategies

Distributed and parallel training strategies are essential for handling the large datasets and complex models typical in generative AI. These strategies involve splitting the training process across multiple processors or machines to speed up computation and improve efficiency. By distributing the workload, these methods enhance performance and contribute to sustainable AI practices by optimizing resource utilization and reducing energy consumption.

Data Parallelism

Data parallelism is one of the most common approaches to distributed training. In this strategy, the dataset is divided into smaller chunks, and these chunks are distributed across multiple processors or machines. Each processor works with a complete copy of the model but trains on its assigned subset of the data. During the training process, each processor performs forward passes on its subset of the data independently and computes gradients locally. These gradients are then aggregated periodically, and the model parameters are updated accordingly. This ensures that all processors maintain a consistent model.

Data parallelism offers several advantages that contribute to sustainable AI practices. First, it is highly scalable, allowing for easy expansion with additional processors or machines. This scalability ensures that large datasets can be processed efficiently, reducing the time and energy required for training. Second, data parallelism is relatively straightforward to implement and understand, making it accessible for a wide range of applications. However, it is important to manage communication overhead, as frequent synchronization of gradients can lead to increased communication costs. Effective batch size management and load balancing are crucial to maximize efficiency and minimize energy consumption.

Model Parallelism

Model parallelism is an alternative approach to distributed training, particularly useful for very large models that cannot fit into the memory of a single device. This strategy divides the model into segments, each assigned to a different processor or machine. Data flows through the partitioned model segments sequentially, with each processor handling its assigned portion of the model. Gradients are computed and back-propagated through the model segments in reverse order, and each processor updates the parameters of its assigned model segment.

Model parallelism offers significant benefits for sustainable AI practices. By enabling the training of very large models that exceed the memory capacity of a single device, it allows for more complex and powerful models to be developed without the need for excessive hardware resources. This reduces the environmental impact associated with hardware production and energy consumption. Additionally, model parallelism reduces communication overhead, as there is less frequent need for parameter synchronization than data parallelism. However, it is essential to ensure an even distribution of computational load across processors and manage the data flow through the partitioned model to minimize idle time and maximize throughput.

Figure 6-2 presents the difference between model and data parallelism. Data parallelism and model parallelism are two distinct approaches to distributed deep learning training. In data parallelism, the input data is split into smaller chunks and processed simultaneously across multiple GPUs, with each GPU containing a full copy of the model. This method is effective for improving training time when the dataset is large. On the other hand, model parallelism involves dividing the layers or parameters of the model itself across multiple GPUs. This approach is particularly useful when dealing with very large models that cannot fit into a single GPU's memory. While data parallelism replicates the entire model on each device, model parallelism shards the model, allowing different parts of it to be processed on different devices simultaneously.

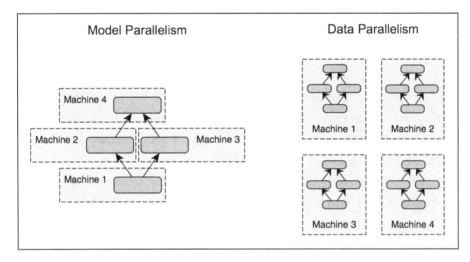

Figure 6-2. *Model and data parallelism*

Hybrid Parallelism and Advanced Strategies

Hybrid parallelism combines elements of both data and model parallelism to leverage the strengths of each approach and optimize resource utilization. This strategy employs data parallelism at a higher level and models parallelism within each data-parallel unit. By balancing intra-model communication with inter-model synchronization, hybrid parallelism maximizes efficiency and reduces energy consumption. In this approach, the overall training process is first split using data parallelism, dividing the dataset among multiple "data-parallel units" or groups of GPUs. Within each data-parallel unit, model parallelism is applied, splitting the model across the GPUs in that unit instead of replicating the entire model on each GPU. This hybrid method allows for processing larger batches of data through data parallelism while simultaneously handling larger models that don't fit on a single GPU through model parallelism. For instance, in a setup with eight GPUs, you might have two data-parallel units, each consisting of four GPUs. The optimal setup though can vary depending on the specific model, dataset, and available hardware. Each unit processes a different subset of the data, and within each unit, the model is split across the 4 GPUs using model parallelism techniques. While this approach can potentially improve computational efficiency, it's worth noting that implementing such hybrid strategies can be complex and requires careful orchestration of communication between GPUs. Specialized frameworks and libraries like DeepSpeed, Megatron-LM, and Alpa are designed to help manage these intricate parallelism strategies more easily

Advanced strategies in distributed and parallel training further enhance sustainable AI practices. Pipeline parallelism divides the model into stages, each assigned to a different processor. Data flows through the pipeline in micro-batches, allowing for concurrent processing of different samples at different stages. This reduces idle time and improves hardware utilization, leading to more efficient energy use. Tensor parallelism splits individual tensors (e.g., weight matrixes) across multiple devices, enabling fine-grained parallelism within model layers. This approach is particularly useful for models with large dense layers, optimizing resource utilization and reducing energy consumption.

The Zero Redundancy Optimizer (ZeRO) eliminates memory redundancies in data-parallel training by sharding model parameters, gradients, and optimizer states across data-parallel processes. This enables training larger models with limited memory resources, reducing the need for excessive hardware and associated energy consumption. Elastic training allows for the dynamic addition or removal of processors during training, enhancing fault tolerance and resource utilization in cloud environments. This flexibility ensures that resources are used efficiently, minimizing energy waste. Figure 6-3 shows how ZeRO transforms data parallelism, boosting memory and computational efficacy.

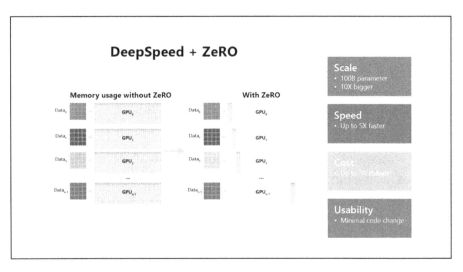

Figure 6-3. ZeRO

Implementation Considerations and Future Directions

Implementing distributed and parallel training strategies requires careful consideration of various factors to ensure sustainable AI practices.

Hardware infrastructure, such as GPU clusters, multinode systems, or cloud-based solutions, should be optimized for high-speed interconnects to facilitate efficient communication. Software frameworks, including distributed training support in popular deep learning frameworks (e.g., PyTorch DistributedDataParallel, TensorFlow Distribution Strategy) and specialized libraries for advanced parallelism (e.g., DeepSpeed, Horovod), play a crucial role in optimizing resource utilization and reducing energy consumption.

Efficient communication protocols, such as all-reduce algorithms for gradient aggregation and NVIDIA Collective Communications Library (NCCL) for GPU-based systems, are essential for minimizing communication overhead and energy use. Fault tolerance mechanisms, including checkpointing and model-saving strategies, help manage node failures and recovery, ensuring that resources are used efficiently. Monitoring and debugging tools, such as distributed logging and performance profiling, are vital for visualizing training progress across multiple nodes and optimizing resource utilization. Figure 6-4 represents how NCCL helps optimize usage of GPUs.

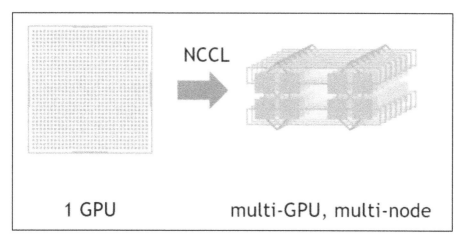

Figure 6-4. *NCCL optimization*

Future directions in distributed and parallel training hold promise for further enhancing sustainable AI practices. Automated parallelism, driven by AI approaches, can determine optimal parallelization strategies and dynamically adapt based on workload and available resources. Federated learning, which involves distributed training across decentralized edge devices while preserving data privacy, can reduce the need for centralized data storage and associated energy consumption. Quantum-accelerated distributed training leverages quantum computing for specific distributed training tasks, potentially reducing energy consumption and improving efficiency. Green AI initiatives focus on developing energy-efficient distributed training strategies to minimize environmental impact. Heterogeneous computing optimizes distributed training across diverse hardware architectures (CPUs, GPUs, TPUs, FPGAs), ensuring that resources are used efficiently and sustainably.

AWS, Azure, and GCP offer a comprehensive suite of tools and services for distributed training and the parallelization of training workloads.

AWS

AWS provides robust distributed and parallel training support through services like Amazon SageMaker and EC2 instances. Amazon SageMaker enables users to distribute training across multiple GPUs with a single click, leveraging instances like the P3, which are powered by NVIDIA V100 Tensor Core GPUs. These instances offer high performance and efficiency, reducing training times from days to minutes. AWS also offers Deep Learning AMIs and Containers, which support popular frameworks such as TensorFlow, PyTorch, and Apache MXNet, allowing users to set up custom environments for domain-specific optimizations. Figure 6-5 presents how the P5 instances and NVIDIA H100 Tensor Core GPUs compare to previous instances and processors.

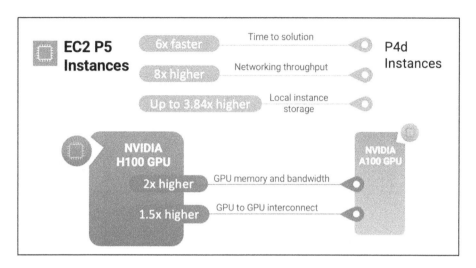

Figure 6-5. Deep learning optimized compute instances

Azure

Azure offers a variety of GPU-enabled virtual machines (VMs), such as the NC-series, ND-series, and NV-series, which are optimized for deep learning and high-performance computing. Azure Machine Learning provides tools for distributed training, enabling users to scale out their training jobs across multiple VMs. Azure's support for frameworks like TensorFlow, PyTorch, and ONNX ensures that users can leverage the best tools for their specific needs. The integration with Azure Batch AI further simplifies distributing training workloads, making it easier to manage and optimize resource utilization.

Figure 6-6 lists the ANC 100 v4-series VMs designed for large-scale AI and machine learning workloads. Powered by NVIDIA A100 Tensor Core GPUs, these VMs provide significant computational power, accelerating model training and inference.

VM Type	GPU type
NC-series	NVIDIA Tesla K80
NCv2-series	NVIDIA Tesla P100
NCv3-series	NVIDIA Tesla V100
NC A100 v4-series	NVIDIA A100 PCIe
NCasT4_v3-series	NVIDIA Tesla T4

Figure 6-6. *ANC 100 v4-series VMs*

GCP

GCP offers GPU-accelerated instances through its Compute Engine, including the A2 instances powered by NVIDIA A100 Tensor Core GPUs. GCP's AI Platform provides managed services for training and deploying machine learning models, with built-in support for distributed training. TensorFlow on GCP, combined with Kubernetes Engine, allows users to easily scale their training jobs across multiple nodes, optimizing performance and resource usage. GCP's Deep Learning VM images come pre-configured with popular frameworks, streamlining the setup process for distributed training.

As the field of AI continues to advance, distributed and parallel training strategies will play an increasingly crucial role in pushing the boundaries of model size, complexity, and performance. Researchers and practitioners must stay abreast of these developments to effectively tackle the challenges of large-scale machine learning and generative AI while promoting sustainable practices. By implementing these strategies and techniques, organizations can significantly enhance the efficiency and sustainability of their generative AI models, ensuring they deliver high performance while minimizing environmental impact.

Optimizing Hyperparameters and Training Schedules

Optimizing hyperparameters and training schedules is critical to developing efficient and high-performing machine learning models. Hyperparameters are settings that govern the training process, such as learning rate, batch size, and the number of layers in a neural network. Properly tuning these hyperparameters can significantly improve model performance and training efficiency. Optimizing hyperparameters in sustainable AI can lead to reduced computational resources and energy consumption, thereby minimizing the environmental impact of AI workloads.

Grid Search

Grid search is a systematic approach to hyperparameter optimization where all possible combinations of hyperparameters are tested. This method involves defining a discrete set of values for each hyperparameter and evaluating the model's performance for every combination of these values. While grid search is exhaustive and can potentially find the optimal hyperparameter settings, it is computationally expensive and time-consuming, especially for large models with many hyperparameters.

Despite its computational cost, grid search can contribute to sustainable model training and inference by ensuring that the selected hyperparameters lead to the most efficient and effective model. By identifying the optimal settings, grid search can minimize the number of training iterations required to achieve desired performance, thereby reducing the overall computational resources and energy consumption. However, due to its exhaustive nature, grid search is often more suitable for smaller models or when computational resources are abundant.

Random Search

Random search offers an alternative to grid search by randomly sampling a fixed number of hyperparameter combinations from the defined search space. Instead of evaluating all possible combinations, random search explores a subset of the hyperparameter space, which can often lead to good results faster than grid search. This method is particularly effective when the search space is large and the complex relationship between hyperparameters and model performance.

Random search contributes to sustainable AI practices by reducing the computational cost associated with hyperparameter optimization. By sampling a limited number of hyperparameter combinations, random search can achieve comparable performance to grid search with significantly fewer evaluations. This reduction in computational effort translates to lower energy consumption and a smaller carbon footprint. Random search's simplicity and ease of implementation make it a practical choice for a wide range of machine learning applications.

Bayesian Optimization

Bayesian optimization is a more sophisticated approach to hyperparameter optimization that uses probabilistic models to predict the performance of different hyperparameter settings. This method focuses on the most promising areas of the hyperparameter space, iteratively refining its predictions based on previous evaluations. Bayesian optimization balances exploration and exploitation, aiming to find the optimal hyperparameters with the fewest evaluations.

Bayesian optimization enhances sustainable model training and inference by efficiently navigating the hyperparameter space and reducing the number of required evaluations. By focusing on the most promising regions, this method can identify optimal hyperparameters faster than grid search or random search, leading to shorter training times and lower energy consumption. Bayesian optimization's ability to handle complex and high-dimensional search spaces makes it particularly valuable for large-scale machine learning projects, where computational resources are a critical concern.

Figure 6-7 compares the three methods.

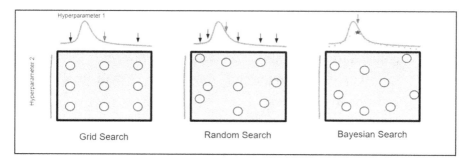

Figure 6-7. *Comparison of grid, random, and Bayesian searches*

Adaptive Learning Rates

Adaptive learning rates are techniques that adjust the learning rate during training to improve convergence speed and model performance. Learning rate schedules, such as step decay, exponential decay, and cosine annealing, systematically reduce the learning rate over time. Adaptive optimizers, such as Adam and RMSprop, dynamically adjust the learning rate based on the gradient information, allowing faster convergence and better performance.

Implementing adaptive learning rates contributes to sustainable AI practices by optimizing the training process and reducing the number of iterations needed to achieve desired performance. By adjusting the learning rate based on the training progress, these techniques prevent the model from getting stuck in suboptimal solutions and ensure efficient use of computational resources. Faster convergence translates to shorter training times and lower energy consumption, making adaptive learning rates a valuable tool for sustainable model training and inference.

Adaptive learning rates offer several advantages over grid search, random search, and Bayesian optimization, particularly in the context of sustainable AI practices. The following points describe why they are often considered superior.

- Dynamic adjustment: Unlike grid, random, or Bayesian searches, which typically set hyperparameters before training begins, adaptive learning rates continuously adjust during the training process. This allows the model to respond to the changing landscape of the loss function in real time.

- Efficiency: Grid, random, and Bayesian searches often require multiple complete training runs with different hyperparameter settings. In contrast, adaptive learning rates optimize within a single training run, significantly reducing computational resources and energy consumption.

- Automatic optimization: Adaptive methods like Adam and RMSprop automatically handle the learning rate optimization, reducing the need for manual tuning. This saves time and computational resources that would otherwise be spent on extensive hyperparameter searches.

- Handling of non-stationary objectives: The loss landscape in deep learning is often non-convex and changes as training progresses. Adaptive methods can handle these changes more effectively than static learning rates determined by grid, random, or Bayesian search.

- Per-parameter adaptation: Many adaptive optimizers (like Adam) adjust learning rates individually for each parameter. This level of granularity is difficult to achieve with traditional hyperparameter search methods.

- Faster convergence: By dynamically adjusting the learning rate, adaptive methods often lead to faster convergence. This directly translates to shorter training times and lower energy consumption.

- Robustness: Adaptive methods are often more robust to initial learning rate choices, making them less sensitive to the initial hyperparameter settings than fixed learning rate schedules.

- Scalability: As models become larger and more complex, the hyperparameter search space grows exponentially. Adaptive methods scale better to these larger models, as they don't require exploring vast hyperparameter spaces.

- Resource efficiency: By optimizing within a single training run, adaptive methods are more resource-efficient, aligning well with sustainable AI practices.

- Generalization: Adaptive methods often lead to solutions that generalize well, potentially reducing the need for extensive fine-tuning or retraining.

While grid, random, and Bayesian searches have their places in hyperparameter optimization, especially for other hyperparameters beyond the learning rate, adaptive learning rates offer a more efficient and effective approach for optimizing the crucial learning rate parameter. This efficiency makes them particularly valuable for sustainable AI practices, where minimizing computational resources and energy consumption is a priority.

AWS, Azure, and GCP also offer a comprehensive suite of services to help with hyperparameter optimization and tuning/training schedules.

AWS

AWS offers several tools for hyperparameter optimization, including Amazon SageMaker's built-in hyperparameter tuning capabilities. SageMaker uses Bayesian optimization to efficiently explore the hyperparameter space, focusing on the most promising areas. This approach reduces the number of required evaluations, leading to faster convergence and lower computational costs. AWS also supports adaptive learning rates through frameworks like TensorFlow and PyTorch, enabling users to implement techniques such as learning rate schedules and adaptive optimizers (e.g., Adam, RMSprop) to improve training efficiency. Figure 6-8 features Amazon SageMaker, which offers managed training and hyperparameter tuning abilities.

Figure 6-8. *Amazon SageMaker*

Azure

Azure Machine Learning provides automated hyperparameter tuning, leveraging Bayesian optimization to find the best hyperparameter settings with minimal evaluations. Azure's support for distributed training allows users to parallelize hyperparameter searches, further reducing the time and resources required. Integrating popular frameworks ensures that users can implement adaptive learning rate techniques to enhance model performance and training efficiency. Azure's comprehensive monitoring and logging tools help users track the progress of their training jobs and make data-driven decisions for optimization. Figure 6-9 highlights Azure's ML capabilities for training and experimentation.

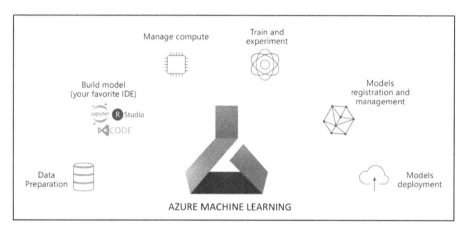

Figure 6-9. *Azure Machine Learning*

GCP

GCP's AI Platform offers hyperparameter tuning services that use Bayesian optimization to efficiently search for the best hyperparameter configurations. This service integrates seamlessly with TensorFlow and other supported frameworks, allowing users to optimize their training schedules and improve model performance. GCP's support for adaptive learning rates and its powerful monitoring and logging capabilities enable users to fine-tune their models and achieve faster convergence with lower computational costs. Figure 6-10 features Vertex AI, a managed ML offering by GCP.

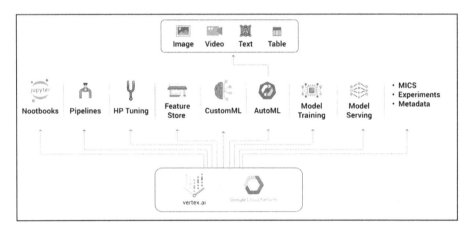

Figure 6-10. *Vertex AI*

Optimizing hyperparameters and training schedules is essential for developing efficient and high-performing machine learning models. Techniques such as grid search, random search, Bayesian optimization, and adaptive learning rates play a crucial role in this process. Each method offers unique advantages and contributes to sustainable AI practices by reducing computational resources and energy consumption. By carefully selecting and tuning hyperparameters, organizations can enhance the efficiency and sustainability of their generative AI models, ensuring they deliver high performance while minimizing their environmental impact.

Efficient Inference Techniques

Inference optimization is a critical aspect of deploying machine learning models, particularly in the context of sustainable AI practices. As models grow in complexity and size, the computational cost and latency associated with running these models become increasingly significant. Efficient model inference techniques focus on reducing these costs, making it possible to deploy models in real-time applications and on resource-constrained devices. By optimizing inference for compute and memory, organizations can significantly enhance the efficiency and sustainability of their generative AI models, ensuring high performance while minimizing environmental impact.

Pruning

Pruning is a technique that involves removing redundant or less important neurons and connections in a neural network. This process is analogous to trimming unnecessary branches from a tree to promote healthier growth. In neural networks, pruning reduces the model size and speeds up inference without significantly impacting performance. The process typically involves identifying and removing connections with small weights or neurons that contribute minimally to the output. While beneficial for reducing model size and improving inference speed, model pruning comes with several drawbacks. It can lead to accuracy loss, especially if pruning is too aggressive, and often requires complex implementation and time-consuming retraining.

Pruned models may become less stable and have a reduced capacity to learn or adapt to new tasks. The technique's effectiveness can vary depending on the model architecture, and finding the optimal pruning level often involves extensive experimentation. Additionally, pruned models may not always translate to faster performance on hardware optimized for dense computations. The potential loss of interpretability and generalization ability are also concerns. These disadvantages underscore the importance of carefully weighing the trade-offs when considering model pruning for a specific application. Figure 6-11 illustrates how model compression can be achieved via pruning.

191

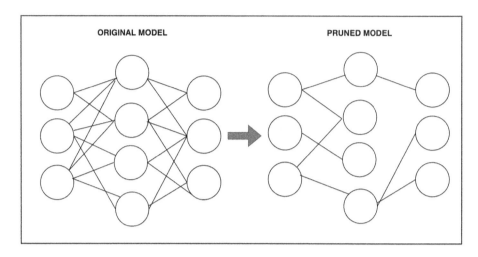

Figure 6-11. *Model pruning*

Pruning contributes to sustainable AI practices in several ways. First, by reducing the model size, pruning decreases the memory footprint of the model, allowing it to run on devices with limited resources. This expanded accessibility means that more efficient, localized processing can occur, reducing the need for energy-intensive data transfers to centralized servers. Second, pruned models require fewer computations during inference, which translates directly to lower energy consumption. This is particularly important for edge devices or mobile applications where battery life is a concern. Finally, pruned models often exhibit improved inference speed, enabling real-time applications that might otherwise be infeasible, thus expanding the potential use cases for AI while maintaining energy efficiency.

Distillation

Knowledge distillation is a process where a smaller, simpler model (referred to as the student) is trained to replicate the behavior of a larger, more complex model (the teacher). The key idea behind distillation is to transfer the knowledge embedded in the complex model to a simpler one, allowing the student model to achieve comparable accuracy while being faster and more efficient. This is typically achieved by training the student model to mimic the teacher model's soft outputs (probabilities) rather than just the hard labels of the training data. Figure 6-12 highlights the core concept that involves initially training a complex network model, referred to as the teacher model. The outputs from this teacher model, along with the true labels of the data, are then used to train a smaller, more efficient network known as the student model. This process, known as knowledge distillation, leverages the teacher model's learned knowledge to enhance the performance of the student model.

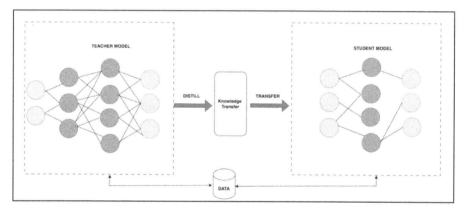

Figure 6-12. *Knowledge distillation*

Distillation offers several benefits for sustainable AI. The most obvious is the reduction in model size and complexity, which leads to lower computational requirements and energy consumption during inference. This is particularly valuable when deploying models on edge devices or in resource-constrained environments. Moreover, distilled models often exhibit improved generalization compared to models of similar size trained directly on the data. This means that smaller, more efficient models can be used without sacrificing significant accuracy, further contributing to energy savings. Additionally, the distillation process can be seen as a form of model compression, allowing organizations to develop and train large, complex models for maximum accuracy and then distill this knowledge into smaller, more deployable versions, optimizing the trade-off between performance and efficiency.

Quantization

Quantization is a technique that converts the model weights from high-precision floating-point numbers (typically 32-bit) to lower-precision representations, such as 8-bit integers. This process significantly reduces the model size and computational requirements, making it particularly suitable for deployment on edge devices or in environments with limited resources. Quantization can be applied post-training or incorporated into the training process (quantization-aware training). Figure 6-13 demonstrates how quantization can help minimize the size of your model while speeding it up for inference at the expense of some accuracy.

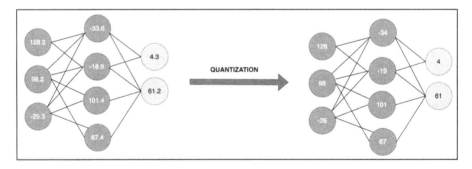

Figure 6-13. *Quantization*

The sustainability benefits of quantization are substantial. By reducing the precision of weights and activations, quantized models require less memory and fewer computational resources. This translates directly to lower energy consumption during inference, as integer operations are generally more energy-efficient than floating-point operations. Furthermore, the reduced model size allows for faster data transfer and loading times, which is particularly beneficial in distributed or edge computing scenarios. Quantization also enables the deployment of complex models on devices with limited processing power or memory, expanding the reach of AI applications without necessitating energy-intensive cloud computing solutions. However, it's important to note that quantization can sometimes lead to a slight decrease in model accuracy, so careful tuning and evaluation are necessary to balance performance and efficiency.

Batching

Batching is a technique that involves aggregating multiple inference requests into a single batch for processing. Instead of processing each input individually, the model processes a group of inputs simultaneously. This approach can significantly improve throughput and resource utilization, especially in high-demand environments where many inference requests are made concurrently. This approach also works when you dont need real time inference and can wait until later to perform processing of data in chunks.

Batching contributes to sustainable AI practices by optimizing the use of computational resources. When processing inputs in batches, the hardware (particularly GPUs) can operate more efficiently, utilizing its parallel processing capabilities to the full extent. This leads to higher throughput and lower energy consumption per inference. In scenarios where the model is deployed in a server environment handling multiple requests, batching can dramatically reduce the overall energy footprint of the inference process. Additionally, batching can help with load balancing and resource allocation, ensuring that computing resources are used optimally and not left idle. However, it's important to balance the benefits of batching with the potential increase in latency for individual requests, especially in real-time applications.

Cloud providers like AWS, Azure, and GCP offer a comprehensive suite of tools and services to optimize inference.

AWS

AWS provides several tools and services to optimize inference, including Amazon Elastic Inference, which allows users to attach just the right amount of GPU acceleration to their EC2 and SageMaker instances. This reduces inference costs by up to 75% compared to using full GPU instances. AWS also supports model pruning and quantization through frameworks like TensorFlow and PyTorch, enabling users to reduce model size and computational requirements. Batching is facilitated by services like AWS Lambda and Amazon SageMaker, which can aggregate multiple inference requests to improve throughput and resource utilization. Figure 6-14 features AWS Inferentia, a high-performance chip designed for inference by Annapurna Labs, now a subsidiary of AWS.

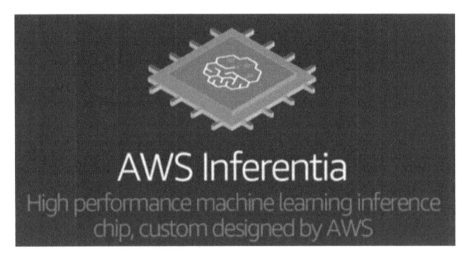

Figure 6-14. *AWS Inferentia*

Azure

Azure offers various services to optimize inference, such as Azure Machine Learning and Azure Functions, which support model deployment and scaling. Azure's support for ONNX Runtime enables efficient inference by optimizing models for different hardware configurations. Techniques like pruning and quantization are supported through frameworks like TensorFlow and PyTorch, allowing users to deploy smaller, more efficient models. Azure's batching capabilities help improve throughput and resource utilization, particularly in high-demand environments.

GCP

GCP provides tools like TensorFlow Lite and TensorFlow Model Optimization Toolkit to facilitate efficient inference on edge devices and resource-constrained environments.

GoogleCloudPlatform/
tensorflow-gcp-tools

Tools that make it easier to run TensorFlow on
Google Cloud Platform (GCP).

Figure 6-15. GCP model optimization tools

These tools support pruning, quantization, and model distillation, enabling users to deploy smaller, faster models without significant loss in accuracy. GCP's AI Platform Prediction service supports batching, allowing users to aggregate multiple inference requests and optimize resource usage. The integration with Kubernetes Engine ensures that inference workloads can be scaled efficiently, reducing latency and computational costs.

Balancing Optimization and Performance

While optimization techniques like pruning, distillation, quantization, and batching offer significant benefits for sustainable AI practices, it's essential to consider the potential trade-offs between optimization and model performance. Striking the right balance is crucial to ensure that the optimized models meet the desired accuracy and performance requirements for specific use cases or applications.

Potential Trade-offs

- Accuracy vs. efficiency: Techniques like pruning and quantization can slightly decrease model accuracy or performance. While these reductions may be negligible in some applications, they can be critical in domains where high precision is essential, such as medical diagnosis or autonomous systems.

- Latency vs. throughput: Batching can improve overall throughput and resource utilization, but it may introduce additional latency for individual inference requests. This trade-off is particularly relevant in real-time applications where low latency is a priority.

- Complexity vs. interpretability: While model compression techniques like distillation can significantly reduce model size and computational requirements, they may also make the resulting models less interpretable or transparent, which can be a concern in domains where model explainability is crucial.

Strategies for Balancing Optimization and Performance

- Adaptive optimization: Develop adaptive optimization frameworks that dynamically adjust the optimization level based on real-time performance metrics or resource availability. It can help balance efficiency and accuracy in dynamic environments or scenarios with varying resource constraints.

- Profiling and benchmarking: Conduct thorough profiling and benchmarking of the optimized models to understand the impact of different techniques on performance metrics like accuracy, latency, and throughput. This data-driven approach can help identify the optimal configuration for specific use cases.

- Incremental optimization: Instead of applying optimization techniques aggressively, consider an incremental approach. Start with a baseline model and gradually apply optimization techniques, monitoring the impact on performance at each step. It allows fine-tuning the optimization level to achieve the desired balance between efficiency and accuracy.

- Hybrid approaches: Explore hybrid approaches that combine multiple optimization techniques. For example, you could apply pruning to remove redundant connections, followed by quantization to reduce precision, and then leverage batching for efficient inference. This layered approach can help maximize each technique's benefits while mitigating their limitations.

- Domain-specific considerations: Tailor the optimization strategies based on the specific requirements and constraints of the target domain or application. For example, accuracy may take precedence over efficiency in safety-critical systems, while in resource-constrained edge devices, efficiency may be the primary concern.

Potential Challenges and Mitigation Strategies

- Optimization overhead: Some optimization techniques, like pruning or distillation, may introduce additional computational overhead during the optimization process. Ensure that the benefits of optimization outweigh the overhead costs by carefully evaluating the trade-offs and considering techniques like quantization-aware training or efficient pruning algorithms.

- Generalization and robustness: Aggressive optimization techniques may impact the model's ability to generalize to unseen data or its robustness to adversarial attacks. Incorporate techniques like adversarial training or data augmentation to enhance the robustness of optimized models.

- Interpretability and transparency: As models become more compressed or distilled, their interpretability and transparency may decrease. Explore techniques like concept activation vectors or layer-wise relevance propagation to maintain interpretability while optimizing models.

- Deployment and integration challenges: Optimized
  models may require specialized hardware or software
  environments for efficient deployment. Ensure
  compatibility with existing infrastructure or plan for
  the necessary upgrades to fully leverage the benefits of
  optimization techniques.

By carefully considering these trade-offs, challenges, and mitigation strategies, organizations can strike the right balance between optimization and performance, ensuring that their generative AI models are efficient, sustainable, and effective for their intended use cases.

Conclusion

Efficient inference techniques play a crucial role in developing sustainable AI systems. Pruning, distillation, quantization, and batching each offer unique advantages in reducing deployed models' computational cost and energy consumption. Pruning and distillation focus on reducing model size and complexity, allowing for more efficient processing. Quantization addresses the precision of computations, enabling deployment on resource-constrained devices. Batching optimizes resource utilization in high-demand scenarios.

By implementing these strategies, organizations can significantly enhance the efficiency and sustainability of their generative AI models. These techniques ensure that models deliver high performance while minimizing their environmental impact. They enable the deployment of AI in a wider range of scenarios, including edge devices and real-time applications, without compromising energy efficiency.

As AI advances, the importance of these efficient inference techniques will only grow. They represent a critical step toward making AI more accessible, efficient, and environmentally friendly. By prioritizing these optimization strategies, the AI community can work toward a future where powerful AI models can be deployed widely and responsibly, with minimal environmental impact.

CHAPTER 7

Cloud and Edge Computing for Generative AI

Cloud computing and edge computing both offer significant potential for enhancing the environmental sustainability of generative AI systems. By consolidating computational resources in centralized, energy-efficient data centers, cloud computing allows using renewable energy sources, advanced cooling technologies, and energy-efficient hardware, thereby minimizing environmental impact. However, the substantial energy requirements of training large language models and other generative AI systems in the cloud highlight the need for more sustainable practices.

Edge computing further enhances sustainability by processing data closer to the source, reducing the need for extensive data transmission to centralized cloud servers. This approach minimizes network traffic and lowers energy consumption associated with data transfer. Power-efficient edge devices with specialized processors enable more efficient AI tasks than traditional data centers. Organizations can achieve a balanced and sustainable computational infrastructure for generative AI systems by combining cloud and edge computing. This hybrid approach allows

© Ishneet Kaur Dua and Parth Girish Patel 2024
I. K. Dua and P. G. Patel, *Optimizing Generative AI Workloads for Sustainability*,
https://doi.org/10.1007/979-8-8688-0917-0_7

resource-intensive training and large-scale inference in energy-efficient cloud data centers. Lighter, fine-tuned models can be deployed on edge devices for localized generative tasks.

Additionally, edge devices can collect and preprocess data, reducing the volume of information sent to the cloud. To maximize the environmental benefits of this hybrid approach, organizations should implement efficient model compression techniques for edge deployment, utilize federated learning to improve models without centralizing data, and design adaptive systems that dynamically shift workloads between cloud and edge based on energy availability and computational demands.

By strategically distributing AI workloads and adopting these practices, organizations can reduce network traffic, optimize resource utilization, and lower the overall carbon footprint of generative AI systems. This approach addresses immediate efficiency needs and promotes a more sustainable digital future, aligning technological advancements with long-term environmental responsibility. Figure 7-1 illustrates how cloud and edge computing are different but often work with each other. Cloud computing focuses on processing data in offsite data centers. On the other hand, edge computing brings the processing closer to the end user in edge devices.

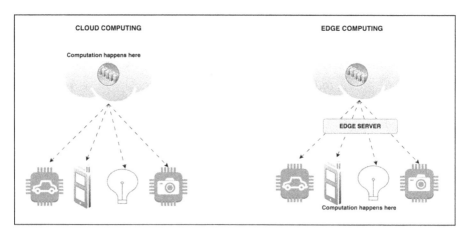

Figure 7-1. *Cloud vs. edge computing*

This chapter walks you through best practices and key architecture patterns associated with edge computing and hybrid infrastructures.

Leveraging Cloud Computing Resources

Cloud computing offers centralized, scalable resources for training and deploying large generative AI models. From a sustainability standpoint, cloud providers can implement energy-efficient data centers and utilize renewable energy sources more effectively than individual organizations.

Major cloud platforms are investing in carbon-neutral or carbon-negative operations, which can help offset the energy consumption of AI workloads. However, the massive energy requirements of training large language models and other generative AI systems in the cloud remain a concern. Researchers estimate that training a single large language model can emit as much CO_2 as five cars over their lifetimes. To address this, cloud providers and AI researchers are exploring more efficient training techniques, such as sparse models and knowledge distillation, to reduce model development's computational and energy costs. One such cloud computing technology is a content delivery network (CDN). Let's dive deeper into that.

How CDNs and Edge Computing Help Environmental Sustainability

Content delivery networks and edge computing are pivotal technologies in the modern digital landscape. They play a crucial role in enhancing the efficiency of data processing and delivery, which can significantly contribute to environmental sustainability. By leveraging the distributed architecture of CDN edge servers and the localized processing capabilities of edge computing, organizations can reduce energy consumption,

205

minimize carbon footprints, and improve overall performance. Figure 7-2 demonstrates how CDN helps improve processing and delivery times.

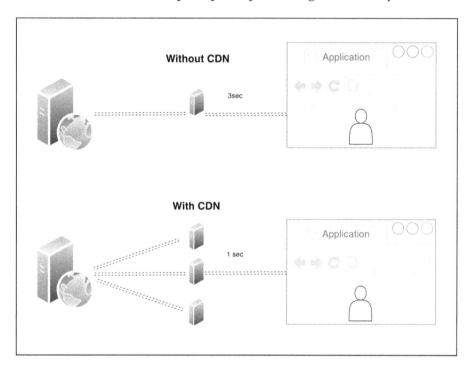

Figure 7-2. *Improvements with CDN*

Distributed Architecture of CDNs

CDNs utilize a network of strategically placed servers to deliver content more efficiently to end users. This distributed architecture means that data can be processed and served from locations closer to the user, reducing the need for long-distance data transfers. By handling and delivering dynamic content at the edge, CDNs minimize the number of trips to the origin server, thereby reducing latency and energy consumption.

Edge Computing for Dynamic Content Delivery

Edge computing enhances the capabilities of CDNs by enabling real-time data processing at the network's edge. This approach allows for personalized and dynamic content delivery without constant communication with central servers. For example, edge servers can process user requests, run AI algorithms, and deliver customized content locally. This reduces the volume of data that needs to be transmitted back and forth, leading to significant energy savings.

Sustainability Benefits of CDNs and Edge Computing

- Reduced data transfer volumes: By processing and caching content closer to the end user, CDNs and edge computing reduce the amount of data that needs to be transmitted over long distances. This decreases energy consumption and lowers the carbon emissions associated with data transfer.

- Lower server loads: Effective caching strategies and localized processing reduce the load on central servers. This means fewer resources are needed to handle peak traffic, resulting in lower energy consumption and a reduced carbon footprint.

- Optimized resource utilization: Continuous monitoring and analytics enable CDNs to operate efficiently. By analyzing traffic patterns and optimizing content delivery routes, CDNs can ensure that resources are used efficiently, reducing energy consumption.

- Enhanced resilience and security: Implementing security measures and multi-CDN strategies can bolster the resilience of content delivery networks. This ensures that services remain available and efficient even during disruptions without compromising sustainability goals.

Best Practices for Sustainable CDN and Edge Computing

- Content optimization: Ensure that content is optimized for delivery, which reduces file sizes and improves load times. This decreases the amount of data that needs to be transferred, saving energy.

- Efficient caching strategies: Implement effective caching strategies to store frequently accessed content at the edge. This reduces the load on the origin servers and minimizes data transfer requirements.

- Leveraging edge computing: Use edge computing to process data locally, reducing the need for long-distance data transmission. This is particularly beneficial for applications requiring real-time processing and low latency.

- Continuous monitoring and analytics: Regularly monitor CDN performance and analyze traffic patterns to identify opportunities for optimization. This ensures that the CDN operates at maximum efficiency, reducing energy consumption.

- Security measures and multi-CDN strategies:
  Implement robust security measures and consider
  using multiple CDNs to enhance resilience. This
  ensures that content delivery remains efficient and
  secure, even during disruptions.

CDNs and edge computing are vital in promoting environmental
sustainability in the digital age. These technologies can significantly
lower energy consumption and carbon emissions by optimizing content
delivery, reducing data transfer volumes, and leveraging localized
processing. Continuous monitoring and implementing best practices
ensure that CDNs operate efficiently, contributing to a more sustainable
digital infrastructure. As organizations increasingly adopt these strategies,
they not only enhance performance and user experience but also align
technological advancements with long-term environmental responsibility.

Edge Computing and On-device AI Optimization

We learned that edge computing brings AI processing closer to where data
is generated, reducing the need for constant data transfer to centralized
cloud servers. This approach can significantly lower energy consumption
and carbon emissions associated with network data transfer.

For generative AI applications, edge computing enables the following.

- Reduced latency for real-time generative tasks

- Enhanced privacy by keeping sensitive data local

- Lower bandwidth usage and associated energy costs

- Improved reliability in areas with limited connectivity

Edge computing and on-device AI optimization represent a transformative shift in how data is processed and managed, offering significant benefits for environmental sustainability. By decentralizing computing power and bringing it closer to the data source, these technologies reduce the energy consumption and carbon footprint associated with traditional, centralized data centers. This section explores how edge computing and on-device AI optimization contribute to sustainable IT, highlighting examples, architectures, case studies, and best practices. However, it's crucial to consider the energy efficiency of edge devices and the environmental impact of manufacturing and disposing of these devices.

Efficient Edge Computing Architectures

Efficient edge computing architectures are crucial in reducing the environmental impact of AI systems. By processing data closer to the source, these architectures minimize the need for extensive data transmission and reduce overall energy consumption. One notable example of efficient edge computing architecture is specialized edge AI chips designed to handle resource-intensive tasks on-device. These chips, such as Qualcomm's Snapdragon processors, are significantly more energy-efficient than traditional data center hardware. For instance, the Snapdragon 8 Gen 3 processor is 30 times more efficient than a data center for image generation tasks. Similarly, the Snapdragon X Elite Compute Platform for laptops demonstrates nearly 28 times more efficiency than running AI tasks in the cloud. These specialized chips enable high-performance AI processing on edge devices while significantly reducing energy consumption and carbon emissions.

We can balance AI advancement and environmental responsibility by processing data closer to the source and leveraging these advanced technologies. As the adoption of these efficient edge computing solutions grows, we can expect a substantial decrease in the overall carbon footprint of AI systems, contributing to broader sustainability goals while still harnessing the power of artificial intelligence.

Hybrid Cloud–Edge Architectures

Hybrid architectures that combine cloud and edge computing offer a balanced approach to generative AI deployment, potentially optimizing both performance and sustainability. This model leverages the strengths of both cloud and edge environments to minimize overall energy consumption and reduce the carbon footprint of AI systems. By strategically distributing AI workloads, hybrid architectures can reduce network traffic, optimize resource utilization, and enhance the efficiency of generative AI applications. Hybrid cloud-edge architectures provide several key benefits for generative AI.

- Resource-intensive training and inference in the cloud: Large-scale AI model training and inference require substantial computational resources. Performing these tasks in energy-efficient cloud data centers allows using advanced hardware, optimized cooling systems, and renewable energy sources, reducing the overall environmental impact.

- Localized generative tasks on edge devices: Lighter, fine-tuned AI models can be deployed to edge devices for localized generative tasks. This reduces the need for constant data transmission to the cloud, lowering bandwidth usage and energy consumption. Edge devices can handle tasks such as real-time image processing, voice recognition, and personalized recommendations, enhancing performance and reducing latency.

- Data collection and preprocessing at the edge: Edge
  devices can collect and preprocess data locally,
  reducing the volume of information sent to the cloud.
  This minimizes network traffic and the associated
  energy costs, making the overall system more efficient
  and sustainable.

Example Architectures

Several architectures exemplify the efficiency of hybrid cloud-edge
computing for generative AI.

- Synergetic big cloud model and small edge models:
  This architecture involves a collaborative framework
  where a large AI model is maintained in the cloud, and
  smaller, task-specific models are deployed on edge
  devices. The cloud handles complex computations and
  model training, while edge devices perform real-time
  inference and data processing. This approach balances
  the computational load and optimizes resource
  utilization across the system.

- Device-centric hybrid AI: In this architecture, the cloud
  offloads AI tasks that edge devices cannot perform
  efficiently. Edge devices run lightweight versions of
  AI models for less complex tasks, while the cloud
  processes more demanding operations. This setup
  allows concurrent processing, where the cloud and
  edge devices work together to deliver accurate and
  timely results.

- Hierarchical edge-cloud architecture: This model introduces intermediate layers between edge devices and the cloud. Local edge servers aggregate data from multiple devices, perform initial processing, and communicate with regional edge centers. These regional centers handle more complex tasks before sending summarized data to the cloud. This tiered approach reduces latency, optimizes bandwidth usage, and allows for more efficient resource allocation.

- Federated learning architecture: In this setup, edge devices collaboratively train a shared model without exchanging raw data. Each device trains on its local data and shares only model updates with a central server in the cloud. The cloud aggregates these updates to improve the global model, which is then redistributed to the edge devices. This architecture preserves privacy and reduces data transfer while leveraging the collective intelligence of distributed devices.

- Dynamic offloading architecture: This flexible architecture allows for real-time decision-making on where to process AI tasks. Based on factors such as device capabilities, network conditions, and energy availability, the system dynamically chooses whether to process a task locally on the edge device or offload it to the cloud. This adaptive approach optimizes performance and energy efficiency across varying conditions.

- Edge-centric generative AI: Lightweight generative models are deployed directly on edge devices in this architecture. These models can perform tasks like text generation, image synthesis, or voice conversion locally. The cloud is used for periodic model updates, complex training, and handling tasks that exceed edge device capabilities. This setup reduces latency for generative tasks and minimizes constant cloud dependence.

- Hybrid transfer learning architecture: This model leverages transfer learning techniques across cloud and edge environments. A base model is trained on a large dataset in the cloud and then transferred to edge devices where it's fine-tuned on local data. This approach reduces the computational burden on edge devices while allowing for personalization and adaptation to local conditions.

- Distributed inference architecture: In this setup, the inference process for complex AI models is distributed across cloud and edge resources. Edge devices handle the initial layers of the model, performing feature extraction and preliminary processing. Intermediate results are then sent to the cloud for final processing through the deeper layers of the model. This distribution balances the computational load and reduces data transfer requirements.

These diverse architectural approaches demonstrate the flexibility and potential of hybrid cloud-edge systems in optimizing generative AI deployment for both performance and sustainability. Each architecture offers unique benefits and can be selected or adapted based on specific use cases, resource constraints, and environmental considerations.

Case Studies

The following case studies demonstrate how hybrid cloud-edge architectures can significantly contribute to sustainability efforts across various sectors. By optimizing data processing and reducing unnecessary data transmission, these systems can lead to substantial energy savings and improved resource management.

- Smart agriculture: In smart agriculture, hybrid architectures monitor and manage crops. Edge devices, such as drones and sensors, collect data on soil moisture, temperature, and plant health. This data is preprocessed locally to identify immediate issues, such as pest infestations or irrigation needs. The processed data is then sent to the cloud for further analysis and long-term planning. This approach reduces the energy consumption associated with data transmission and enhances the efficiency of agricultural operations. Figure 7-3 classifies the process for smart agriculture to manage farms using modern information and communication technologies to increase the quantity and quality of products while optimizing the human labor required.

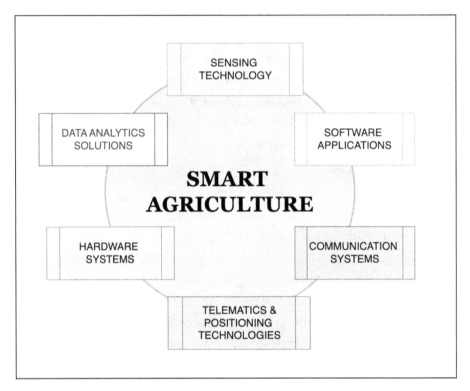

Figure 7-3. *Smart agriculture*

- Healthcare: In healthcare, hybrid architectures enable real-time monitoring and diagnostics. Wearable devices and sensors collect patient data, such as heart rate and blood pressure, and process it locally to detect anomalies. Critical data is then transmitted to cloud-based systems for comprehensive analysis and storage. This reduces the latency in detecting health issues and minimizes the energy costs of continuous data transmission.

- Smart cities: A major European city implemented a hybrid cloud-edge architecture for its smart city initiatives. Edge devices were deployed throughout the city to collect traffic, air quality, and energy consumption data. Local processing at the edge allowed for real-time traffic management and air quality alerts. The cloud component aggregated data from all edge nodes for long-term urban planning and sustainability analysis. This approach reduced data center energy consumption by 30% compared to a centralized cloud-only solution. See Figure 7-4.

Figure 7-4. Smart cities

- Renewable energy management: A large wind farm operator implemented a hybrid architecture to optimize energy production and maintenance. Edge devices on individual turbines process real-time data on wind speed, direction, and equipment performance. This local processing allows for immediate adjustments to turbine operations. Aggregated data is sent to the cloud for predictive maintenance and overall farm optimization. The system reduced unnecessary data transmission by 60%, leading to significant energy savings in network infrastructure.

- Sustainable manufacturing: A global manufacturing company deployed a hybrid cloud-edge system across its factories. Edge computing nodes in each facility process real-time production data, enabling immediate adjustments to optimize energy use and reduce waste. This approach led to a 25% reduction in energy consumption and a 15% decrease in material waste across the company's operations. The infographic shown in Figure 7-5 demonstrates how the cloud component analyzes data from all facilities to identify global efficiency improvements and what goes into the process.

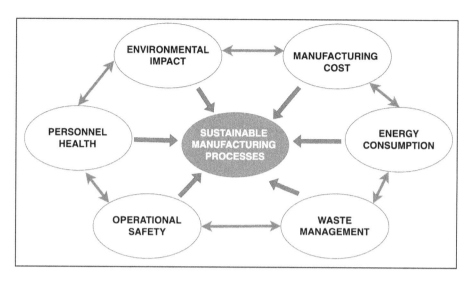

Figure 7-5. *Sustainable manufacturing process*

- Electric vehicle charging network: A national electric vehicle charging network utilizes a hybrid architecture to manage its charging stations. Edge computing at each station optimizes local charging schedules based on real-time grid conditions and vehicle needs. The cloud component manages overall network load balancing and predicts future charging demands. This system improved energy efficiency by 20% and reduced peak load on the power grid by 15%.

- Sustainable retail: A large retail chain implemented a hybrid cloud-edge system for its stores. Edge devices in each store manage inventory, adjust lighting and HVAC based on occupancy, and process point-of-sale data. The cloud component analyzes data from all stores to optimize supply chain logistics and energy use across the network. This approach reduced energy consumption in stores by 18% and decreased transportation-related carbon emissions by 12%.

- Water management: A water utility company deployed a hybrid architecture to manage its distribution network. Edge devices at pumping stations and key points in the network process data on water flow, pressure, and quality. Local processing allows for immediate leak detection and pressure adjustments. The cloud component analyzes long-term trends and optimizes overall network efficiency. This system reduced water loss by 25% and energy consumption for water distribution by 20%.

Best Practices for Efficient Hybrid Architectures

Consider the following best practices to maximize the environmental benefits of hybrid cloud-edge architectures for generative AI.

- Model compression techniques: Implementing efficient model compression techniques, such as pruning, quantization, and knowledge distillation, can reduce the size and complexity of AI models. This makes them more suitable for deployment on edge devices, enhancing their efficiency and reducing energy consumption.

- Federated learning: Federated learning allows AI models to be trained across multiple edge devices without centralizing the data. This approach reduces the need for data transmission to the cloud, lowering energy consumption and enhancing data privacy. Each device processes local data and shares model updates, minimizing network traffic.

- Adaptive workload distribution: Designing adaptive systems that dynamically shift workloads between cloud and edge environments based on energy availability and computational demands can optimize resource utilization. For instance, non-critical tasks can be processed in the cloud during off-peak hours, while time-sensitive tasks are handled at the edge, ensuring efficient energy use.

Hybrid cloud-edge architectures offer a promising pathway to achieving sustainable IT by reducing energy consumption and minimizing the carbon footprint of generative AI systems. By leveraging the strengths of both cloud and edge computing, organizations can optimize performance, enhance efficiency, and reduce environmental impact. As the adoption of hybrid architectures continues to grow, they will play a crucial role in driving environmental sustainability in the era of advanced AI and digital transformation.

Edge Computing Tools by Major Cloud Providers

Edge computing has become an integral part of modern cloud architectures, enabling data processing and analysis closer to the source of data generation. Major cloud providers have developed a range of tools and services to support edge computing initiatives. These tools enhance the environmental sustainability of generative AI by processing data closer to the source, reducing the need for extensive data transfer and network usage. This localized processing lowers power consumption and optimizes resource utilization by running smaller, efficient AI models at the edge. Real-time processing and adaptive power management minimize energy use, while federated learning across decentralized devices reduces reliance on massive data centers. Additionally, edge computing supports offline capabilities and hardware-specific optimizations, collectively leading to more energy-efficient and environmentally friendly generative AI applications.

Amazon Web Services

Figure 7-6 identifies the edge computing offerings from Amazon Web Services (AWS), focusing on their key features and applications.

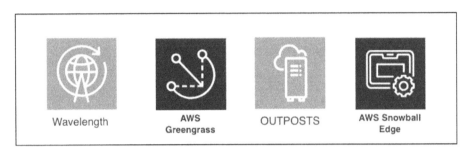

Figure 7-6. *Edge computing services*

- AWS IoT Greengrass: This Internet of Things service extends AWS capabilities to edge devices, allowing them to act locally on the data they generate while still using the cloud for management, analytics, and durable storage. Greengrass enables edge devices to run AWS Lambda functions, execute predictions based on machine learning models, keep device data in sync, and communicate with other devices securely.

- AWS Outposts: A fully managed service that extends AWS infrastructure, services, APIs, and tools to virtually any data center, co-location space, or on-premises facility. It enables a consistent hybrid experience by allowing customers to run computing and storage on-premises while seamlessly connecting to AWS's broad array of services in the cloud.

- AWS Snowball Edge: A data migration and edge computing device with on-board storage and compute power. It can be used to move terabytes to petabytes of data into and out of AWS, as a temporary storage tier for large local datasets, or to support local workloads in remote or offline locations.

- AWS Wavelength: This service embeds AWS compute and storage services within 5G networks, providing mobile edge computing infrastructure for developing, deploying, and scaling ultra-low-latency applications.

Microsoft Azure

Figure 7-7 presents the edge computing offerings from Microsoft Azure, focusing on their key features and applications.

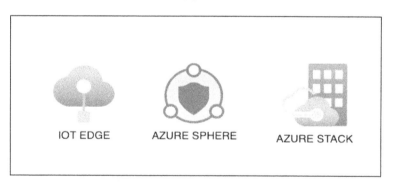

IOT EDGE   AZURE SPHERE   AZURE STACK

Figure 7-7. *Azure edge stack*

- Azure IoT Edge: This Internet of Things service that builds on top of IoT Hub. It allows for deploying and running artificial intelligence, Azure services, and custom logic directly on IoT devices. Azure IoT Edge enables edge devices to run in offline or intermittent connection scenarios, reducing IoT solution costs and ensuring local data residency.

- Azure Stack: A portfolio of products that extend Azure services and capabilities to your environment of choice—from the data center to edge locations and remote offices. It includes Azure Stack Hub for on-premises environments, Azure Stack HCI for virtualized infrastructures, and Azure Stack Edge for edge computing scenarios.

- Azure Sphere: A comprehensive IoT security solution that includes hardware (Azure Sphere-certified microcontrollers), software (Azure Sphere OS), and cloud components (Azure Sphere Security Service) for IoT device security.

- Azure Percept: A comprehensive, easy-to-use platform for creating edge AI solutions. It includes hardware accelerators integrated with Azure AI and IoT services, making it easier to build and deploy AI-powered apps at the edge.

Google Cloud Platform

Figure 7-8 shows the edge computing offerings from Google Cloud Platform (GCP).

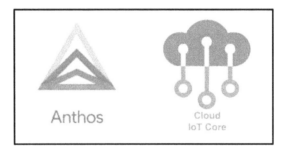

Figure 7-8. GCP edge stack

- Anthos: A managed application platform that extends Google Cloud services and engineering practices to edge environments. It enables consistent application deployment and management across on-premises, edge, and multiple public clouds.

- Cloud IoT Core: A fully managed service to connect, manage, and ingest data from millions of globally dispersed devices. When used with other Google Cloud services, it provides a complete solution for collecting, processing, analyzing, and visualizing IoT data in real time.

- Cloud IoT Edge: A software stack that extends Google Cloud's data processing and machine learning capabilities to edge devices. It consists of two main components: the Edge TPU (a hardware accelerator) and the Edge TPU runtime, which allows TensorFlow Lite models to be executed on the Edge TPU.

- Google Distributed Cloud Edge: A fully managed hardware and software solution that brings Google Cloud's infrastructure and services closer to where data is being generated and consumed.

IBM Cloud

Figure 7-9 displays the edge computing offerings from IBM Cloud.

Figure 7-9. *IBM edge stack*

- IBM Edge Application Manager: A comprehensive edge computing platform that helps enterprises deploy, manage, and monitor applications and workloads at the edge across multiple edge sites and thousands of edge devices.

- IBM Cloud Pak for Applications: A hybrid cloud solution that helps accelerate the development of applications built for Kubernetes. It can modernize existing applications and develop new cloud-native apps for deployment at the edge.

- IBM Watson Anywhere: This service allows organizations to run IBM Watson AI models and applications in any cloud or on-premises environment, including edge locations.

- IBM Maximo Application Suite: An integrated platform that uses AI, IoT, and analytics to optimize asset performance. It includes edge computing capabilities for real-time data processing and analysis.

Oracle Cloud

Figure 7-10 presents the edge computing offerings from OCI.

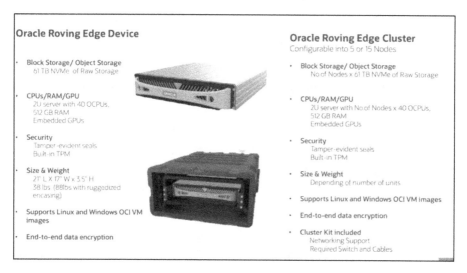

Figure 7-10. *OCI edge stack*

- Oracle Cloud VMware Solution: This service provides a dedicated, cloud-native VMware-based environment that enables customers to migrate and run VMware workloads on Oracle Cloud Infrastructure, including at the edge.

- Oracle Roving Edge Infrastructure: A mobile, ruggedized edge computing device that can be deployed in remote and harsh environments. It brings core infrastructure services to the edge, enabling low-latency, high-bandwidth access to cloud applications and services.

- Oracle IoT Cloud: A platform that enables organizations to connect, analyze, and integrate data from IoT devices. It includes edge computing capabilities for processing data closer to the source.

Major cloud providers have developed a wide array of tools and services to support edge computing initiatives. These offerings range from IoT platforms and edge devices to fully managed edge infrastructure solutions. As edge computing continues to evolve, we expect these providers to further enhance their edge capabilities, enabling organizations to build more sophisticated and efficient distributed applications.

The choice of edge computing tools depends on various factors, including existing cloud infrastructure, specific use cases, and required features. Organizations should carefully evaluate these offerings to determine which best aligns with their edge computing strategy and overall IT architecture. As edge computing becomes increasingly important in the era of IoT, 5G, and AI, these tools will play a crucial role in enabling organizations to process data closer to the source, reduce latency, and build more responsive and efficient applications. The continued development and refinement of these edge computing tools by major cloud providers will be essential in driving innovation and addressing the growing demands of edge computing across various industries.

Conclusion

The future of hybrid cloud-edge architectures for environmentally sustainable generative AI lies in continuous innovation and research. The following are key areas of focus.

- Advanced model compression: Developing more sophisticated model compression techniques to further reduce the computational requirements of AI models without sacrificing performance.

- Edge AI hardware: Advancing the capabilities of edge AI hardware to handle more complex tasks efficiently. This includes the development of specialized processors and low-power AI chips.

- Collaborative intelligence: Enhancing the collaboration between cloud and edge systems to improve the overall efficiency and effectiveness of AI applications. This involves developing new algorithms and frameworks for distributed AI processing.

- Sustainable practices: Implementing sustainable practices in designing, deploying, and operating hybrid cloud-edge systems. This includes using renewable energy sources, optimizing resource utilization, and minimizing waste.

- Enhanced security and privacy: Developing robust security and privacy measures tailored to hybrid cloud-edge environments. This includes implementing advanced encryption techniques, secure multiparty computation, and privacy-preserving machine learning methods to protect sensitive data and ensure compliance with regulations.

- Edge-to-cloud continuum: Creating a seamless edge-to-cloud continuum where data and workloads can move fluidly between edge and cloud based on real-time needs and conditions. This involves designing adaptive systems that dynamically allocate resources and optimize performance across the spectrum.

- AI-driven optimization: Leveraging AI itself to optimize hybrid cloud-edge architectures. AI-driven tools can predict workload patterns, optimize resource allocation, and identify energy-saving opportunities, thereby enhancing the overall efficiency and sustainability of the system.

- Scalability and interoperability: Ensuring that hybrid cloud-edge architectures can scale effectively and interoperate with various technologies and platforms. This includes developing standardized protocols and APIs that facilitate seamless integration and scalability across different environments.

- Real-time analytics and decision-making: Enhancing the capabilities of edge devices to perform real-time analytics and decision-making. This includes developing lightweight, real-time AI models that can operate efficiently on edge hardware, enabling immediate insights and actions.

- User-centric design: Focusing on user-centric design principles to ensure that hybrid cloud-edge systems are intuitive, accessible, and tailored to the needs of end users. This includes designing user-friendly interfaces and providing tools that allow users to easily manage and optimize their AI workloads.

In conclusion, the environmental sustainability of generative AI systems depends on the thoughtful implementation of cloud and edge computing strategies. By leveraging the strengths of each approach and adopting energy-efficient practices, organizations can harness the power of generative AI while minimizing its ecological impact. As the field evolves, continued research into sustainable AI practices and hardware innovations will be crucial for balancing technological advancement with environmental responsibility.

CHAPTER 8

Energy-Efficient AI Deployment and Scaling

Efficient AI Deployment and Scaling

In the rapidly evolving landscape of artificial intelligence (AI) and machine learning (ML), efficient deployment and scaling of applications have become crucial challenges. As AI workloads grow in complexity and resource demands, organizations are increasingly turning to containerization and virtualization strategies to optimize their infrastructure and streamline AI operations. These technologies offer powerful solutions for managing the unique requirements of AI applications, from development and testing to production deployment and scaling.

This chapter delves into the world of containerization and virtualization, exploring their fundamental concepts, benefits, and specific applications in the context of AI workloads. We examine how these strategies contribute to efficient resource utilization, enhanced portability, improved scalability, and better isolation of AI applications. Additionally, we discuss best practices, tools, and techniques for implementing these strategies effectively in AI environments.

© Ishneet Kaur Dua and Parth Girish Patel 2024
I. K. Dua and P. G. Patel, *Optimizing Generative AI Workloads for Sustainability*,
https://doi.org/10.1007/979-8-8688-0917-0_8

Containerization and virtualization are important for environmental sustainability as they enable more efficient use of computational resources, reducing the energy consumption and carbon footprint of IT operations. By optimizing resource utilization, these technologies help minimize the need for additional physical hardware, thereby decreasing electronic waste and the environmental impact associated with manufacturing and disposing of IT equipment. Furthermore, they facilitate the use of cloud computing, which often relies on energy-efficient data centers powered by renewable energy sources. This shift not only enhances the scalability and flexibility of AI operations but also aligns with global efforts to promote sustainable practices in technology and reduce the environmental impact of digital infrastructure.

Understanding Containerization

Containerization is a lightweight virtualization technology that allows applications and their dependencies to be packaged together in a standardized unit called a container. Unlike traditional virtual machines, containers share the host operating system's kernel, resulting in minimal overhead and faster startup times.

Key Features of Containerization

Containers offer significant advantages in software deployment and management. They encapsulate an application along with its dependencies, ensuring consistent behavior across various environments, which enhances portability. With minimal overhead, containers allow for a higher density of applications to run on a single host, contributing to efficiency. Their ability to be started and stopped rapidly enables quick scaling and updates, making deployment processes more agile. Additionally, each container operates in its own isolated environment, which enhances security and minimizes conflicts between applications. Finally, container images can be versioned, facilitating easier rollbacks and updates, which further streamline application management.

Popular Containerization Technologies

- Docker: The most widely used containerization platform, Docker provides tools for building, running, and managing containers.

- containerd: An industry-standard container runtime that is used by Docker and other container orchestration platforms.

- CRI-O: A lightweight container runtime specifically designed for Kubernetes.

- Podman: An alternative to Docker that doesn't require a daemon and can be run rootless.

These technologies are featured in Figure 8-1.

Figure 8-1. *Popular containerization technologies*

Containerization in AI Workflows

Containerization brings numerous advantages to AI workflows, enhancing the development, deployment, and management of artificial intelligence applications. By ensuring consistency across development, testing, and production stages, containers promote reproducibility, which is crucial for maintaining the integrity of AI models. The complex ecosystem of AI libraries and their dependencies becomes more manageable through containerization, as these components can be easily packaged and controlled within a single container. Scalability is significantly improved, allowing containerized AI applications to expand horizontally to meet increased demand. This approach also optimizes resource utilization, which is particularly beneficial for GPU-accelerated AI workloads that require efficient hardware management. Additionally, containerization facilitates version control of AI models, enabling teams to maintain and deploy different iterations using container images, thus streamlining the development and update process.

Virtualization is a technology that enables multiple virtual instances of operating systems or applications to run on a single physical machine by abstracting the underlying hardware. This abstraction allows for more efficient use of resources and greater flexibility in managing workloads. Key features of virtualization include resource abstraction, which separates logical resources from physical hardware for flexible allocation, and isolation, where each virtual machine operates in its own secure environment. Virtual machines also benefit from hardware independence, allowing them to run on different types of hardware, which enhances portability. Additionally, virtualization supports snapshots and cloning, making it easy to back up, clone, or roll back virtual machines to previous states. Resource overcommitment is another advantage, allowing the hypervisor to allocate resources (like memory, CPU, etc.) more dynamically and efficiently, leading to better overall utilization of physical resources.

There are several types of virtualization: full virtualization, which provides a complete simulation of the underlying hardware for unmodified guest operating systems; paravirtualization, which requires modifications to the guest operating system (OS) for better performance; hardware-assisted virtualization, which uses CPU features to enhance performance; and OS-level virtualization, or containerization, where the OS kernel supports multiple isolated user-space instances. Popular virtualization technologies, as seen in Figure 8-2, include VMware vSphere, an enterprise-grade platform with advanced features for data centers; Kernel-based Virtual Machine (KVM), an open-source technology integrated into the Linux kernel; Xen, an open-source hypervisor used in cloud computing; and Microsoft Hyper-V, a platform integrated with Windows Server.

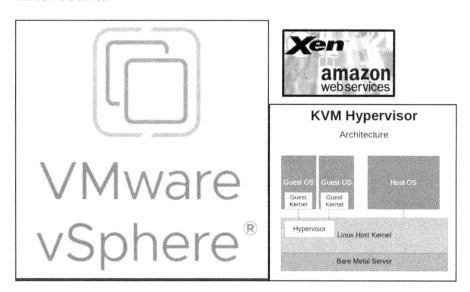

Figure 8-2. *Popular virtualization technologies*

Virtualization in AI Workflows

Virtualization offers several benefits for AI workflows by providing strong resource isolation, ensuring that resource-intensive AI tasks do not interfere with other applications. It allows for flexibility by enabling different operating systems or environments to run on the same hardware, supporting diverse AI development and testing scenarios. A virtual machine (VM) can be easily backed up or rolled back, facilitating experimentation and testing of AI models. Additionally, virtualization abstracts hardware, making it easier to move AI workloads between different physical machines or cloud environments. The isolation provided by VMs enhances security for sensitive AI workloads and data.

When comparing containerization and virtualization for AI workloads, each has distinct advantages. Containerization is lightweight, with minimal overhead, allowing for a higher density of AI applications. Containers start almost instantly, enabling rapid scaling of AI services, and they share the host OS kernel, reducing resource consumption. Containers are also highly portable across different environments. On the other hand, virtualization provides stronger isolation, which is crucial for sensitive AI workloads. Virtual machines can run different operating systems on the same host and emulate different hardware configurations. Virtualization platforms often have more mature management and monitoring tools.

Choosing between containerization and virtualization for AI workloads depends on several factors. For high-performance AI applications, containers may be preferred due to lower overhead. If strong isolation is required, virtual machines might be a better choice. For highly portable AI applications, containers are often the better option. Containers are generally more efficient for maximizing resource utilization, especially in cloud environments. If the AI workload requires specific operating systems or hardware configurations, virtual machines might be necessary. Containers often integrate better with modern DevOps practices and CI/CD pipelines, making them suitable for development workflows.

Both containerization and virtualization contribute significantly to environmental sustainability by optimizing the use of computational resources. Containerization, with its lightweight nature and efficient resource utilization, reduces energy consumption and carbon footprint associated with running AI workloads. This efficiency minimizes the need for additional physical hardware, thereby decreasing electronic waste and the environmental impact of manufacturing and disposing of IT equipment. Similarly, virtualization allows multiple virtual machines to run on a single physical server, maximizing resource utilization and reducing the overall energy consumption of data centers. Both technologies facilitate the use of cloud computing, which often relies on energy-efficient data centers powered by renewable energy sources. This shift enhances the scalability and flexibility of AI operations while promoting sustainable practices in technology and reducing the environmental impact of digital infrastructure.

Hybrid Approaches

In many cases, such as the following, a hybrid approach combining containerization and virtualization can provide the best of both worlds.

- Containerized applications on VMs: Running containerized AI applications inside VMs can provide both the efficiency of containers and the strong isolation of VMs.

- Nested virtualization: Some virtualization platforms support running containers or even other VMs inside a VM, offering multiple layers of abstraction and isolation.

- Orchestration across VMs and containers: Tools like Kubernetes can manage workloads across both VMs and containers, providing a unified management plane.

Optimizing Resource Utilization and Isolation

Efficient resource utilization and proper isolation are crucial for AI workloads, which often demand significant computational resources and may handle sensitive data. Both containerization and virtualization offer strategies to optimize these aspects. Dynamic resource allocation tools can automatically adjust allocations based on workload demands. Setting appropriate resource quotas and limits prevents resource hogging and ensures fair distribution. GPU sharing technologies allow multiple AI workloads to efficiently share GPU resources, and memory overcommitment in virtualization platforms can maximize resource utilization, though caution is needed with memory-intensive AI workloads. For performance-critical AI tasks, CPU pinning can dedicate specific CPU cores to containers or VMs.

Isolation techniques are equally important. Network policies or virtual networks can isolate AI workloads and control traffic flow, while dedicated storage volumes or virtual disks isolate data for different AI workloads. Containerization's process isolation prevents interference between different AI applications. In cloud environments, security groups control access to AI workloads, and for highly sensitive tasks, trusted execution environments provide an additional layer of security.

Resource scheduling plays a vital role in optimizing AI workloads. Affinity and anti-affinity rules control the placement of AI workloads based on resource availability, performance requirements, or data locality. Priority classes ensure critical AI workloads get resources ahead of less important tasks. In Kubernetes environments, taints and tolerations control which nodes can run specific AI workloads. Custom schedulers can be implemented to account for specific resource requirements of AI workloads, such as GPU availability or memory bandwidth. Advanced scheduling algorithms can optimize the placement of AI workloads for maximum resource utilization.

As AI technology advances and requires more computational power, effectively utilizing containerization and virtualization becomes increasingly crucial. Organizations that implement best practices and continuously refine their strategies can build adaptable, efficient, and scalable environments for AI workloads. This approach fosters innovation and maintains a competitive edge in the AI-driven landscape.

Monitoring and continuous optimization are vital in this context because they enable organizations to help with the following tasks.

- Performance monitoring: Implement comprehensive resource usage, application performance, and system health monitoring for AI workloads.

- Predictive scaling: Use machine learning models to proactively predict resource requirements and scale.

- Continuous optimization: Regularly analyze resource usage patterns and adjust allocations to optimize efficiency.

- Cost monitoring: In cloud environments, implement cost monitoring and optimization tools to ensure efficient use of resources.

- Given the complexity of most IT ecosystems, some companies may have concerns about how best to get started with continuous implementation. To help simplify the process and make it more accessible, continuous monitoring can be seen as a flywheel where we perform ongoing overall and risk assessment, monitor system definitions, and use the right set of software tools, as seen in Figure 8-3.

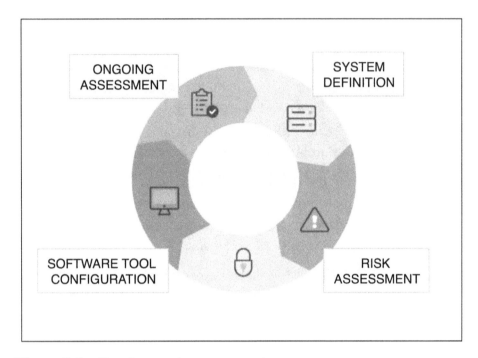

Figure 8-3. *Continuous improvement process*

To optimize containerization and virtualization for AI workloads, several best practices should be implemented. Careful analysis of resource requirements is essential for right-sizing containers or VMs, ensuring efficient allocation. Creating optimized container images or VM templates specifically for AI workloads, including only necessary components, can significantly improve performance. Strict version control for these images and templates is crucial to maintain the reproducibility of AI environments. Security hardening through regular patching, access controls, and encryption is vital for protecting sensitive AI data and models.

Performance tuning of container runtimes and hypervisors, considering factors like CPU governor settings, NUMA awareness, and I/O optimization, can enhance AI workload efficiency. Integrating containerization and virtualization into CI/CD pipelines enables automated testing and deployment of AI models. Efficient data management strategies, including considerations for data locality, caching, and storage performance, are critical for AI workloads.

Designing containerized or virtualized AI applications with scalability in mind, encompassing both horizontal and vertical scaling options, ensures adaptability to changing demands. Implementing robust backup and disaster recovery strategies safeguards containerized and virtualized AI environments. Finally, aligning containerization and virtualization strategies with regulatory requirements and organizational governance policies ensures compliance and proper management of AI workloads.

Containerization and virtualization strategies offer powerful tools for efficient AI deployment and scaling. By leveraging these technologies, organizations can optimize resource utilization, enhance portability, improve scalability, and ensure proper isolation of AI workloads. The choice between containerization and virtualization, or a hybrid approach, depends on the specific requirements of the AI application and the overall infrastructure strategy.

Cloud and Edge Computing Strategies

Cloud Computing Strategies for Energy-Efficient AI Deployments

Cloud computing has become a cornerstone for deploying AI applications due to its scalability, flexibility, and cost-effectiveness. For energy-efficient AI deployments, several strategies can be employed.

Resource Optimization

Cloud platforms offer tools and services to optimize resource utilization, such as autoscaling, which adjusts the number of active servers based on current demand. This ensures that resources are not wasted, reducing energy consumption and costs.

Green Data Centers

Many cloud providers are investing in green data centers powered by renewable energy sources. By choosing cloud services that prioritize sustainability, organizations can significantly reduce the carbon footprint of their AI deployments. Providers like Google Cloud Platform, Amazon Web Services (AWS), and Microsoft Azure have committed to carbon neutrality and are increasingly using solar, wind, and hydropower to run their operations.

Efficient Data Management

Efficient data management practices, such as data compression, deduplication, and tiered storage, can reduce the energy required for data processing and storage. By minimizing the volume of data that needs to be processed, these strategies help conserve energy and improve overall system performance.

Edge Computing and Its Implications for Energy Efficiency in Generative AI

Edge computing brings computation and data storage closer to the location where it is needed, which can significantly enhance energy efficiency for AI applications, particularly those involving generative AI.

Reduced Latency and Bandwidth

By processing data closer to the source, edge computing reduces the need to transfer large volumes of data to centralized cloud servers. This not only decreases latency, improving real-time processing capabilities, but also reduces bandwidth usage, which can lead to energy savings.

Localized Processing

Edge devices, such as IoT devices and edge servers, can perform localized processing of AI models, which reduces the load on central data centers. This distributed approach to computation can lead to significant energy savings, especially for applications that require constant data analysis and decision-making, such as autonomous vehicles and smart cities.

Energy-Efficient Hardware

Edge computing often utilizes specialized, energy-efficient hardware designed for low-power environments. Devices such as Raspberry Pi, NVIDIA Jetson, and Intel Movidius are examples of hardware that can perform AI inference at the edge with minimal energy consumption, making them ideal for sustainable deployments.

Smaller Language Models in Cloud and Edge Computing

Small language models (SLMs) present a promising approach for deploying AI efficiently in both cloud and edge environments. These models are designed to be lightweight, requiring fewer computational resources and less energy compared to their larger counterparts. Some of those models are shown in Figure 8-4.

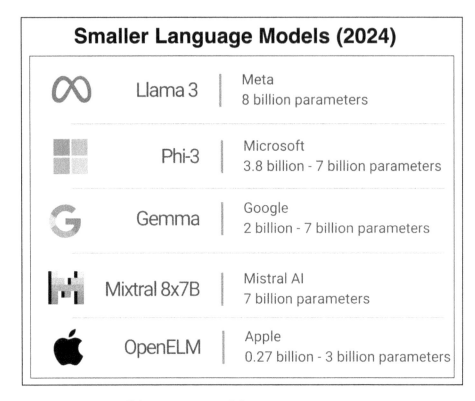

Figure 8-4. *Small language models*

Benefits of Smaller Language Models

- Reduced computational load: SLMs require significantly fewer parameters and computational operations, making them ideal for deployment on resource-constrained devices such as edge servers and IoT devices. This reduction in computational load translates directly to lower energy consumption.

- Faster inference: Their smaller size allows SLMs to perform inference more quickly than larger models. This is particularly beneficial in edge computing scenarios where real-time processing is critical, such as in autonomous systems and real-time analytics.

- Lower latency: With SLMs, data can be processed locally on edge devices, reducing the need for data to travel to and from centralized cloud servers. This not only decreases latency but also conserves bandwidth and energy.

Deployment Strategies for Smaller Language Models

- Model pruning and quantization: Techniques such as model pruning (removing unnecessary parameters) and quantization (reducing the precision of model weights) can be employed to create SLMs without significantly compromising performance. These techniques help in reducing the model size and computational requirements, making them suitable for edge deployment.

- Federated learning: Federated learning allows SLMs to be trained across multiple decentralized devices without sharing raw data. This approach not only enhances privacy but also reduces the energy costs associated with data transfer and centralized training.

- Hybrid cloud-edge architectures: In hybrid architectures, critical tasks can be handled by SLMs on edge devices, while more complex processing can be offloaded to the cloud. This balance ensures that energy-intensive computations are minimized at the edge while still leveraging the cloud's computational power when necessary.

Autoscaling and Load Balancing Techniques

In the rapidly evolving landscape of generative AI, efficiently managing and scaling resources is crucial for maintaining optimal performance and cost-effectiveness. Autoscaling and load balancing techniques play a pivotal role in achieving these goals, enabling AI systems to dynamically adjust to varying workloads and distribute tasks efficiently across available resources. See Figure 8-5.

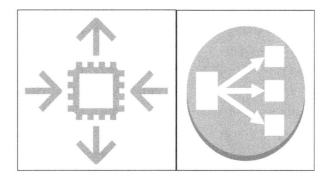

Figure 8-5. *Autoscaling groups and load balancing*

Autoscaling for Generative AI Workloads

Autoscaling is a critical feature for generative AI systems, allowing them to automatically adjust the number of active instances based on current demand. This capability is particularly important for generative AI workloads, which often experience significant fluctuations in demand due to factors such as time of day, user activity patterns, or sudden spikes in requests.

Horizontal Scaling

Horizontal scaling, also known as scaling out, involves adding or removing instances of a service based on demand. This approach is well-suited for generative AI workloads that can be parallelized across multiple instances.

Kubernetes Horizontal Pod Autoscaler (HPA) is a popular tool for implementing horizontal scaling in containerized environments.

The HPA works by monitoring specified metrics, such as CPU utilization or custom application-specific metrics, and automatically adjusting the number of pods in a deployment or replica set. For generative AI workloads, custom metrics might include the number of pending requests, model inference time, or GPU utilization.

Implementing horizontal scaling for generative AI workloads requires careful consideration of several factors.

- Metric selection: Choosing the right metrics to trigger scaling events is crucial. While CPU utilization is a common metric, it may not always accurately reflect the load on AI models. Custom metrics that directly measure model performance or queue length can provide more accurate scaling triggers.

- Scaling thresholds: Setting appropriate thresholds for scaling up or down is essential to avoid unnecessary oscillations. For generative AI workloads, these thresholds might need to be more aggressive to handle sudden spikes in demand.

- Cooldown periods: Implementing cooldown periods between scaling events helps prevent rapid fluctuations and stabilizes the system after a scaling action.

- Resource allocation: Ensuring that each instance has sufficient resources to handle incoming requests is crucial. This may involve careful tuning of resource requests and limits in Kubernetes.

- Statelessness: Designing generative AI services to be stateless facilitates horizontal scaling, as new instances can be added or removed without concern for maintaining state.

Vertical Scaling

Vertical scaling, or scaling up, involves adjusting the resource allocation (CPU, memory, GPU) of a single instance. This approach can benefit generative AI workloads that require more resources to handle complex tasks or larger models.

Kubernetes Vertical Pod Autoscaler (VPA) is a powerful tool for automatically adjusting CPU and memory resource requests of pods based on their usage patterns, which can be particularly beneficial for generative AI workloads. This approach is especially useful in scenarios where model complexity varies, requiring different resource allocations for specific models and for batch processing tasks that need temporary resource increases without the overhead of creating new instances. In resource-constrained environments, vertical scaling can optimize the use of existing resources. Implementing vertical scaling for generative AI workloads involves comprehensive resource monitoring to inform scaling decisions, configuring the VPA to make gradual adjustments to avoid disrupting running workloads, setting appropriate upper and lower bounds for resource allocation to prevent over or under-provisioning, and integrating with horizontal scaling strategies for a more flexible and efficient scaling solution. By leveraging vertical scaling, organizations can ensure their generative AI workloads receive the necessary resources dynamically, improving overall performance and resource utilization.

Load Balancing for Efficient Resource Provisioning

Load balancing is essential for distributing incoming requests evenly across available resources, preventing any single resource from becoming a bottleneck. For generative AI workloads, effective load balancing can significantly impact performance, latency, and resource utilization.

Load balancing techniques play a crucial role in distributing generative AI workloads efficiently across available resources. Round Robin, a simple method that distributes requests sequentially among instances, is straightforward to implement and works well for homogeneous workloads. However, it may not be optimal for generative AI tasks with varying complexities or durations.

The Least Connections method, which directs traffic to the instance with the fewest active connections, can be more effective for generative AI workloads with varying completion times, helping to prevent the overloading of specific instances. However, it may not account for actual resource utilization. Resource-based load balancing, which allocates requests based on current resource utilization, is particularly effective for generative AI workloads as it considers the actual resource consumption of AI tasks. This method adapts well to workloads with varying resource requirements and can handle heterogeneous instances, although it requires more complex implementation and monitoring. Each of these methods has its own pros and cons and can be implemented using various tools such as Kubernetes Services, cloud load balancers, or custom schedulers, depending on the specific requirements of the Generative AI application.

Advanced Load Balancing Techniques for Generative AI

Beyond the basic methods, several advanced load balancing techniques can be particularly beneficial for generative AI workloads.

- Content-based routing: Routes requests based on the content or characteristics of the AI task

  - Can direct specific types of AI generation tasks to specialized instances or models

- Weighted load balancing: Assigns different weights to instances based on their capabilities

  - Useful when dealing with heterogeneous hardware (e.g., different GPU types)

- Adaptive load balancing: Dynamically adjusts load balancing algorithms based on observed performance metrics

  - Can optimize for factors like latency, throughput, or resource efficiency

- Session affinity: Ensures that requests from the same user or session are directed to the same instance

  - Can be beneficial for maintaining context in conversational AI applications

- Predictive load balancing: Uses machine learning models to predict future load and preemptively adjust distribution

  - Can help smooth out the impact of sudden spikes in demand

Implementing Effective Autoscaling and Load Balancing for Generative AI

Several best practices should be considered to implement effective autoscaling and load balancing for generative AI workloads. Comprehensive monitoring is essential, involving detailed tracking of both system-level metrics (CPU, memory, GPU utilization) and application-specific metrics (request queue length, model inference time) using tools like Prometheus and Grafana for real-time visibility. Developing and exposing custom metrics that accurately reflect the performance and resource utilization of AI models is crucial, and these metrics should be integrated with autoscaling and load balancing solutions. Gradual scaling policies should be implemented to avoid sudden large changes in capacity, utilizing step scaling or target tracking scaling policies in cloud environments. Preemptive scaling based on historical patterns or upcoming events ensures capacity is available before demand spikes occur. Setting appropriate resource quotas and limits helps prevent runaway scaling and control costs, with regular reviews and adjustments based on observed usage patterns. Thorough load testing is necessary to understand the scaling characteristics of AI workloads, and this information should be used to fine-tune autoscaling and load balancing configurations.

Combining horizontal and vertical scaling strategies provides maximum flexibility, using vertical scaling for fine-grained adjustments and horizontal scaling for larger capacity changes. Implementing intelligent load shedding mechanisms allows for graceful handling of scenarios where demand exceeds maximum capacity, such as queuing less critical requests or providing degraded service for non-essential features. For global applications, geo-distributed load balancing routes requests to the nearest available resources, reducing latency and improving user experience. Continuous optimization involves regularly analyzing performance data and adjusting autoscaling and load balancing configurations, potentially using AI-driven optimization tools to refine resource allocation strategies continuously.

Challenges and Considerations

While autoscaling and load balancing can significantly enhance the efficiency of generative AI workloads, several challenges and considerations must be addressed. Cold start times for large GenAI models can be substantial, necessitating strategies like maintaining warm pools of instances or using lightweight proxy models for initial responses. Effective state management is crucial, especially for applications that maintain state; this may involve using distributed caching or session replication techniques. Cost management is another important factor, as autoscaling can lead to unexpected cost increases; setting up alerts and automated actions can help mitigate this risk. Robust model versioning and deployment strategies, such as blue-green deployment or canary releases, ensure consistency across scaled instances. Data locality should be considered for workloads requiring access to large datasets, with caching strategies or data replication implemented to ensure efficient data access. For GPU-accelerated workloads, intelligent GPU scheduling and sharing mechanisms, like NVIDIA MPS, are necessary. Finally, compliance and data privacy must be ensured, with appropriate data isolation and access controls implemented across scaled instances to meet relevant data protection regulations.

Future Trends in Autoscaling and Load Balancing for Generative AI

As generative AI continues to evolve, several trends are expected to shape the future of autoscaling and load balancing in this domain. AI-driven optimization will see increased use of AI and machine learning techniques to make autoscaling and load balancing decisions, leading to self-tuning systems that adapt to changing workload patterns without human intervention. The growth of serverless platforms specifically designed for AI workloads will offer fine-grained scaling and billing, integrating GenAI capabilities into existing serverless platforms. Autoscaling and load balancing techniques optimized for edge computing scenarios will develop, balancing workloads between edge devices and cloud resources based on latency and bandwidth constraints. As quantum computing advances, scaling and load balancing strategies will leverage both classical and quantum resources, facilitating hybrid classical-quantum workload distribution for applicable generative AI tasks. There will also be an increased focus on energy-efficient scaling strategies to reduce the environmental impact of AI workloads, integrating carbon-aware scheduling and load balancing techniques.

Autoscaling and load balancing are critical for the efficient deployment and operation of GenAI workloads. By implementing a combination of horizontal and vertical scaling strategies, along with intelligent load balancing techniques, organizations can ensure optimal performance, resource utilization, and cost-effectiveness of their AI systems. As GenAI advances, autoscaling and load balancing strategies must evolve to meet the increasing demands of more complex models and diverse application scenarios. Continuous monitoring, optimization, and adaptation of these strategies will be key to maintaining efficient and responsive GenAI systems amid ever-changing workloads and technological advancements.

Monitoring and Optimizing Energy Consumption in Generative AI Deployments

As generative AI models continue to grow in size and complexity, their energy consumption has become a significant concern for both environmental sustainability and operational costs. Monitoring and optimizing energy consumption in generative AI deployments is crucial for organizations looking to balance performance with energy efficiency. This section explores various techniques for monitoring energy consumption and strategies for energy-aware scheduling and resource allocation in generative AI systems.

Techniques for Monitoring Energy Consumption

Effective energy management begins with accurate monitoring. In GenAI deployments, this involves tracking the power usage of hardware components and the efficiency of software processes. Hardware-based power meters are essential for measuring energy consumption at various levels, including rack-level, server-level, and component-level, providing insights into power usage and helping to establish baseline consumption patterns, identify energy-intensive components, track optimization efforts, and detect anomalies. Software tools like Intel Power Gadget, NVIDIA Management Library (NVML), AMD µProf, and various Linux-based tools offer detailed insights into the power usage of specific components, particularly CPUs and GPUs. Cloud provider tools such as AWS CloudWatch, Google Cloud Monitoring, and Azure Monitor also provide energy metrics and efficiency recommendations.

Integrated energy dashboards aggregate data from various sources to offer a comprehensive view of energy consumption, featuring real-time monitoring, trend visualization, alerts for abnormal patterns, and reporting capabilities. Application-level energy profiling, which includes techniques like code instrumentation and energy-aware profilers, helps identify energy hotspots within AI models and enables energy-aware optimization. Thermal monitoring, using infrared thermal imaging and temperature sensors, provides additional insights into energy efficiency by highlighting heat distribution and cooling performance.

Network and storage systems, often overlooked, also contribute significantly to overall energy consumption. Monitoring the energy usage of network devices and storage systems can identify energy-intensive operations. Long-term energy consumption analysis, involving historical data collection, trend analysis, and correlation with workload patterns, is crucial for understanding energy usage patterns, informing capacity planning, predicting future energy needs, and enabling accurate budgeting and forecasting for energy expenses.

Conclusion

Optimizing the lifecycle of generative AI for sustainability involves a holistic approach that includes containerization and virtualization for efficient deployment, autoscaling and load balancing for dynamic resource management, and continuous monitoring and optimization of energy consumption. By implementing these strategies, organizations can achieve significant improvements in both performance and sustainability, contributing to the responsible development and deployment of AI technologies.

CHAPTER 9

Sustainable AI Lifecycle Management

Sustainable AI lifecycle management is an emerging discipline that addresses the environmental, social, and economic impacts of artificial intelligence throughout its lifecycle. As AI technologies become increasingly integrated into various sectors, it is imperative to adopt practices that ensure their development, deployment, and maintenance are sustainable. This chapter explores the principles and strategies for managing AI systems in a way that minimizes their carbon footprint, promotes ethical use, and ensures long-term viability. By examining the entire lifecycle—from data collection and model training to deployment and decommissioning—this chapter aims to provide a comprehensive framework for integrating sustainability into AI practices, which fosters innovation that aligns with global sustainability goals.

Model Retraining and Updating Strategies

Discuss Strategies for Efficient Model Retraining and Updating in Generative AI

In the rapidly evolving field of generative AI, maintaining the accuracy and relevance of models is critical. Efficient model retraining and updating strategies are essential to ensure that AI systems continue performing well

© Ishneet Kaur Dua and Parth Girish Patel 2024
I. K. Dua and P. G. Patel, *Optimizing Generative AI Workloads for Sustainability*,
https://doi.org/10.1007/979-8-8688-0917-0_9

as new data becomes available and the underlying data distribution shifts. This section explores various strategies to achieve this objective.

Scheduled Retraining

Scheduled retraining is a systematic approach to updating machine learning models at predetermined intervals. This strategy involves retraining the model on a regular basis, such as weekly, monthly, or quarterly, depending on the specific needs of the application and the rate of change in the underlying data. The primary purpose of scheduled retraining is to ensure that the model remains accurate and relevant over time by incorporating new data and adapting to changes in the data distribution. This approach addresses two common challenges in machine learning.

- Data drift refers to changes in the statistical properties of the input data over time.

- Concept drift refers to shifts in the relationships between input features and the target variable.

- Concept drift and data drift are two significant challenges in maintaining the performance of machine learning models over time. Concept drift occurs when the relationship between input features and the target variable changes, affecting the model's accuracy. Examples include evolving credit scoring conditions, changing spam email characteristics, and shifting product preferences.

  In contrast, data drift happens when the statistical properties of input variables change without necessarily altering the underlying relationships. This can be seen in scenarios such as sensor degradation, population demographic shifts, image quality changes, and economic indicators fluctuations. Both types of drift can significantly impact model performance, necessitating ongoing monitoring and periodic adjustments to ensure continued accuracy and relevance of machine learning models in real-world applications.

By regularly updating the model with fresh data, scheduled retraining helps maintain its performance and reliability in dynamic environments.

The following are some best practices for implementing scheduled retraining.

- Optimal retraining frequency: Analyze the rate of change in your data and model performance to establish an appropriate retraining schedule. This may involve monitoring performance metrics and conducting regular data analysis.

- Retraining process automation: Implement automated pipelines that handle data preparation, model retraining, and evaluation to ensure consistency and reduce manual effort.

- Version control: Maintain clear versioning of models, datasets, and code to track changes and enable rollbacks if necessary.

- Performance monitoring: Continuously monitor model performance to identify any degradation that may require adjustments to the retraining schedule.

- Data quality checks: Implement robust data validation processes to ensure that new data used for retraining meets quality standards.

- Resource optimization: Schedule retraining during off-peak hours to minimize disruption to production systems and optimize resource utilization.

- Incremental learning: When possible, use incremental learning techniques to update the model with new data without retraining from scratch, reducing computational costs.

- Alternative strategies: Regularly evaluate and assess whether scheduled retraining is the most effective approach for your use case, considering alternatives like online learning or event-driven retraining.

- Documentation: Maintain clear documentation of the retraining process, including any hyperparameter tuning or architectural changes made during each iteration.

- Compliance and governance: Ensure that the retraining process adheres to relevant regulatory requirements and organizational policies, particularly in sensitive domains.

While scheduled retraining can be resource-intensive, as it often involves retraining the entire model from scratch, these best practices can help organizations implement this strategy effectively and efficiently, ensuring that their machine learning models remain accurate and reliable over time.

Event-driven Retraining

Event-driven retraining is an adaptive approach to updating machine learning models based on specific triggers or events rather than at fixed intervals. This strategy involves monitoring the model's performance and the data environment and initiating a retraining process when certain predefined conditions are met.

In event-driven retraining, the model is updated in response to specific triggers, which may include significant changes in the input data distribution, performance degradation beyond a certain threshold, availability of a substantial amount of new data, detection of concept drift or data drift and external events that may impact the model's relevance (e.g., regulatory changes, market shifts).

This approach can be more efficient than scheduled retraining because it focuses on updating the model only when necessary, which conserves computational resources and ensures the model remains relevant to current conditions.

The following are some best practices for implementing event-driven retraining.

- Defined triggers: Establish specific, measurable criteria for initiating the retraining process. These could be based on performance metrics, data volume thresholds, or statistical tests for distribution shifts.

- Robust monitoring: Set up and implement a comprehensive monitoring system that tracks model performance, data distributions, and other relevant metrics in real-time or near-real-time.

- Retraining pipeline: Develop an automated system to trigger and execute the retraining process when the defined conditions are met, minimizing manual intervention.

- Triggers: Regularly review and validate the effectiveness of your triggers to ensure they're accurately identifying situations that warrant retraining.

- Safeguards: Implement checks and balances to prevent unnecessary or too frequent retraining, which could lead to overfitting or resource waste.

- Version control: Maintain clear versioning of models, datasets, and code to track changes and enable rollbacks if necessary.

- Performance evaluation: Implement a robust evaluation framework to compare the performance of the retrained model against the current production model before deployment.

- Gradual rollout: Use techniques like canary deployments or A/B testing to gradually introduce the retrained model and monitor its performance in real-world conditions.

- Resource management: Ensure that your infrastructure can handle sudden spikes in computational demand that may occur during event-driven retraining.

- Documentation and alerting: Maintain detailed logs of retraining events, including the triggers that initiated them, and set up alerting systems to notify relevant team members when retraining occurs.

- Continuous improvement: Regularly review the effectiveness of your event-driven retraining strategy and refine it based on observed outcomes and changing business needs.

By implementing these best practices, organizations can leverage event-driven retraining to maintain model accuracy and relevance while optimizing resource usage and adapting quickly to changes in the data environment.

Active Learning

Active learning is a strategic approach in machine learning that enhances the efficiency of model retraining by selectively querying the most informative data points for labeling. Instead of relying on a large, randomly sampled dataset, the model identifies and requests labels for data points that are expected to provide the greatest improvement in its performance. This targeted approach minimizes the amount of labeled data required, making it particularly valuable in scenarios where labeled data is scarce or expensive.

Moreover, active learning contributes to environmental sustainability by reducing the computational resources needed for training machine learning models. By focusing only on the most informative data points, active learning decreases the overall data processing and storage requirements, leading to lower energy consumption. This efficiency not only speeds up the training process but also reduces the carbon footprint associated with extensive data handling and computation, aligning with efforts to promote sustainable practices in technology.

Active learning operates on the principle that not all data points are equally valuable for training a model. By focusing on the most informative data points, the model can achieve higher accuracy with fewer labeled examples. The process typically involves the following steps.

1. Initial model training: Start with a small, labeled dataset to train an initial version of the model.

2. Query strategy: Use a query strategy to identify the most informative data points from a larger pool of unlabeled data. Common strategies include uncertainty sampling, query-by-committee, and expected model change.

3. Labeling: Request labels for the selected data points from human annotators or other labeling sources.

4. Model update: Retrain the model using the newly labeled data points.

5. Iteration: Repeat the process iteratively, continually improving the model's performance with each cycle.

The following are some best practices for active learning.

- Choose an appropriate query strategy. Select a query strategy that aligns with your specific use case and data characteristics. For example, uncertainty sampling selects data points where the model is least confident, while query-by-committee involves multiple models to identify data points with the most disagreement.

- Start with a representative initial dataset. Ensure that the initial labeled dataset is representative of the overall data distribution. This helps the model make more accurate initial predictions and improves the effectiveness of subsequent active learning cycles.

- Automate the query process. Implement automated systems to handle the selection and querying of data points. This reduces manual effort and speeds up the active learning process.

- Integrate human-in-the-loop. Use human annotators to label the selected data points. Ensure that annotators are well-trained and understand the labeling guidelines to maintain consistency and accuracy.

- Monitor model performance. Continuously monitor the model's performance to assess the impact of the newly labeled data. Adjust the query strategy if necessary to ensure optimal improvement.

- Evaluate cost-benefit ratios. Regularly evaluate the cost-benefit ratio of active learning. Ensure that the reduction in labeled data requirements justifies the additional complexity and computational resources involved.

- Scalability: Design the active learning system to scale with the volume of data and the number of iterations. This includes efficient data storage, processing capabilities, and annotation workflows.

- Establish a feedback loop. Incorporate insights from the active learning process into future iterations. This helps refine the query strategy and improve overall model performance.

- Use diverse query strategies. Consider using multiple query strategies in tandem to capture different aspects of data informativeness. This can lead to a more robust and well-rounded model.

- Recognize ethical considerations. Ensure that the active learning process adheres to ethical guidelines, particularly when dealing with sensitive or personal data. Maintain transparency and fairness in the selection and labeling of data points.

By following these best practices, organizations can effectively implement active learning to enhance the efficiency of model retraining. This approach reduces the need for extensive labeled datasets and accelerates the development of high-performing machine learning models, making it a valuable strategy in various applications.

Transfer Learning

Transfer learning is a machine learning technique that leverages knowledge gained from solving one problem to improve performance on a different but related task. In this approach, a model trained on a large dataset for a specific task is repurposed or fine-tuned to perform well on a new, often smaller dataset or related task. The key idea behind transfer learning is that the features and patterns learned by a model on one task can be useful for other tasks, especially when the source and target domains share similarities. This approach is particularly valuable when dealing with limited labeled data for the target task or when aiming to reduce the computational resources and time required for training.

In terms of environmental sustainability, transfer learning is beneficial because it reduces the need for extensive data collection and processing, which can be resource-intensive. By reusing existing models and knowledge, transfer learning minimizes the computational power required to train new models from scratch, thereby lowering energy consumption and the associated carbon footprint. This efficiency not only accelerates the development of machine learning applications but also supports sustainable practices by conserving resources and reducing the environmental impact of AI technologies.

There are several ways to implement transfer learning.

- Feature extraction: The pre-trained model is a fixed feature extractor, and only the final layers are retrained on the new task.

- Fine-tuning: Some or all layers of the pre-trained model are updated using the new data, allowing the model to adapt to the specific characteristics of the target task.

- Domain adaptation: The model is adjusted to perform well on a new domain that is related but distinct from the original training domain.

The following are some best practices for transfer learning.

- Appropriate source model: Select a pre-trained model that was trained on a large, diverse dataset relevant to your target task.

- Source and target tasks: Transfer learning is most effective when the source and target tasks share similar features or patterns.

- Selectively: Depending on the size of your target dataset and its similarity to the source task, decide whether to freeze early layers (feature extraction) or fine-tune the entire model.

- Learning rates: Adjust lower learning rates for pre-trained layers and higher rates for newly added layers to preserve learned features while allowing adaptation.

- Data augmentation: Enhance the diversity of your target dataset through data augmentation techniques to improve generalization.

- Regularization: Apply regularization techniques like weight decay or dropout to prevent overfitting, especially when fine-tuning on a small dataset.

- Performance monitoring: Continuously evaluate the model's performance on both the source and target tasks to ensure positive transfer and avoid negative transfer.

- Iterative refinement: Start with a simple transfer learning approach and iteratively refine it based on performance results.

- Layer freezing: Experiment with freezing different combinations of layers to find the optimal balance between preserving learned features and adapting to the new task.

- Multitask learning: If applicable, train the model on multiple related tasks simultaneously to improve overall performance and generalization.

By following these best practices, organizations can effectively leverage transfer learning to build high-performing models with less data and computational resources, accelerating the development of machine learning solutions across various domains.

Ensemble Learning

Ensemble learning is a machine learning technique that enhances model performance by combining the predictions of multiple models. These models, often called "base learners" or "weak learners," are trained on different subsets of the data or using various algorithms. The final output is derived by aggregating the predictions of these individual models through methods such as averaging, voting, or stacking. The primary advantage of ensemble learning is its ability to improve the robustness and accuracy of the model. By leveraging the strengths of multiple models, ensemble methods can mitigate the weaknesses of individual models, leading to better overall performance. Additionally, when updating the model, only a subset of the ensemble may need to be retrained, rather than the entire ensemble, which reduces the computational burden.

Ensemble learning also contributes to environmental sustainability by optimizing the efficiency of machine learning processes. By improving model accuracy and reducing the need for retraining large models, ensemble techniques can decrease the computational resources required, thus lowering energy consumption. This efficiency helps minimize the carbon footprint associated with extensive data processing and model training. Furthermore, ensemble learning can be applied to environmental monitoring and prediction tasks, such as assessing the impact of climate change or predicting ecological phenomena, thereby supporting sustainable resource management and decision-making.

The following are some best practices for implementing ensemble learning.

- Diverse base learners: Use a variety of algorithms or training data subsets to ensure diversity among the base learners. This diversity helps capture different aspects of the data and reduces the risk of overfitting.

- Balanced complexity: Ensure that the base learners are not overly complex, which can lead to overfitting. Combining simpler models can often yield better generalization.

- Aggregation method: Choose an appropriate method for combining predictions. Common techniques include majority voting for classification tasks and averaging for regression tasks. More sophisticated methods like stacking can also be used, where a meta-model is trained to combine the predictions of the base learners.

- Cross-validation: Use cross-validation to evaluate the performance of the ensemble. This helps ensure that the ensemble method generalizes well to unseen data.

- Incremental updating: When updating the ensemble, retrain only the base learners that show significant performance degradation. This approach conserves computational resources and allows for more efficient model updates.

- Hyperparameter tuning: Optimize the hyperparameters of both the base learners and the ensemble method. Proper tuning can significantly enhance the performance of the ensemble.

- Regularization: Apply regularization techniques to prevent overfitting, especially when combining multiple complex models. Techniques such as dropout or weight decay can be beneficial.

- Model interpretability: While ensembles can be more accurate, they are often less interpretable than single models. Use techniques like feature importance analysis or SHAP values to understand the contributions of individual models and features.

- Scalability: Design the ensemble learning system to scale with the volume of data and the number of base learners. Efficient data processing and parallel computing can help manage large-scale ensembles.

- Continuous monitoring: Continuously monitor the performance of the ensemble model in production. Set up alerts for performance degradation and automate the retraining process to maintain model accuracy over time.

Organizations can implement ensemble learning to build robust and accurate machine learning models by following these best practices. This approach improves model performance and provides a flexible and efficient framework for model updating and maintenance.

Online Learning

Online learning in machine learning is a technique where the model is continuously updated as new data arrives rather than being trained on a fixed dataset. This approach allows the model to adapt in real time to changes in the data distribution, making it particularly useful for applications with streaming data or rapidly changing environments. In online learning, the model is updated incrementally with each new data point or small batch of data rather than being retrained from scratch on the entire dataset. This enables the model to learn from new information quickly and efficiently without the need for storing and processing large amounts of historical data.

Online learning also contributes to environmental sustainability by reducing the computational resources required for model training. Since the model is updated incrementally, it avoids the need for extensive retraining sessions that consume significant energy and computational power. This efficiency leads to lower energy consumption and a reduced carbon footprint. Additionally, by processing data in real-time, online learning can be applied to environmental monitoring systems, enabling timely responses to ecological changes and supporting sustainable resource management. This adaptability not only enhances the model's performance but also aligns with efforts to promote sustainable practices in technology.

The following are some of the key characteristics of online learning.

- Continuous adaptation: The model evolves as new data becomes available, allowing it to capture changing patterns and trends.

- Memory efficiency: Online learning algorithms typically require less memory, as they don't need to store the entire dataset.

- Real-time processing: Models can be updated in real-time, making them suitable for applications that require immediate responses to new data.

- Handling concept drift: Online learning can adapt to changes in the underlying data distribution, known as concept drift.

The following are some best practices for implementing online learning.

- Choose appropriate algorithms. Select online learning algorithms that are suitable for your specific problem and data characteristics. Examples include stochastic gradient descent, online passive-aggressive algorithms, and follow-the-regularized-leader.

- Manage learning rate. Use adaptive learning rate techniques to balance between quick adaptation and stability. Methods like AdaGrad or Adam can be effective.

- Implement regularization. Apply regularization techniques to prevent overfitting, especially when dealing with high-dimensional data or noisy streams.

- Handle concept drift. Implement mechanisms to detect and adapt to concept drift, such as using sliding windows or adaptive model ensembles.

- Maintain a validation set. Keep a separate validation set to monitor the model's performance and detect potential issues like overfitting or underfitting.

- Consider mini-batch updates. For efficiency, update the model with small batches of data rather than individual instances, especially in high-throughput scenarios.

- Implement data preprocessing. Ensure that incoming data is properly preprocessed and normalized before being used to update the model.

- Monitor model stability. Regularly assess the model's stability and performance to ensure it's not becoming unstable or diverging over time.

- Implement safeguards. Set up mechanisms to prevent the model from being adversely affected by outliers or malicious data points.

- Balance exploration and exploitation. In some applications, it may be necessary to balance between exploiting the current model and exploring new patterns in the data.

- Consider hybrid approaches. Combine online learning with periodic batch updates or ensemble methods to leverage the strengths of different approaches.

- Implement logging and versioning. Maintain detailed logs of model updates and implement versioning to track changes and enable rollbacks if necessary.

By following these best practices, organizations can effectively implement online learning to build adaptive and efficient machine learning models that can handle streaming data and evolving environments. This approach enables continuous learning and adaptation, making it valuable for a wide range of applications in dynamic and real-time scenarios.

Model Distillation

Model distillation is a technique used to create a smaller, more efficient model that retains the performance of a larger, more complex model. In this process, a "teacher" model, which is the larger model, is used to train a "student" model, the smaller model, by transferring its knowledge. This approach is particularly useful for updating models in resource-constrained environments, as the distilled model requires less computational power and memory.

Model distillation contributes to environmental sustainability by significantly reducing the computational resources needed to deploy machine learning models. By creating smaller models that maintain high performance, model distillation decreases the energy consumption associated with running and maintaining AI systems. This reduction in resource usage leads to a lower carbon footprint, making AI technologies more environmentally friendly. Additionally, the efficiency gained through model distillation can be particularly beneficial in deploying AI solutions for environmental monitoring and management, enabling more sustainable and scalable applications in areas such as climate modeling and resource conservation.

The following are some best practices for implementing model distillation.

- Choose an appropriate teacher model. Select a high-performing, well-trained teacher model that captures the necessary knowledge and patterns in the data.

- Design an effective student model. Create a student model that is significantly smaller and less complex than the teacher model but still capable of learning the essential information.

- Use soft targets. During the distillation process, use the soft targets (probabilistic outputs) from the teacher model rather than hard labels. This helps the student model learn the nuanced patterns captured by the teacher.

- Optimize the loss function. Combine the standard loss function with a distillation loss that measures the difference between the student model's outputs and the teacher model's soft targets. Adjust the weighting of these components to balance learning from the teacher and the actual data.

- Experiment with temperature scaling. Apply temperature scaling to the teacher model's output probabilities to smooth the distribution and make it easier for the student model to learn. Experiment with different temperature values to find the optimal setting.

- Train incrementally. Train the student model incrementally, starting with a lower learning rate and gradually increasing it as the model converges. This helps stabilize the training process and improves convergence.

- Use data augmentation. Data augmentation techniques provide diverse training examples, helping the student model generalize better and capture the essential knowledge from the teacher model.

- Apply regularization techniques. Regularization techniques such as dropout or weight decay can prevent overfitting and improve the generalization ability of the student model.

- Evaluate performance. Continuously monitor and evaluate the performance of the student model against the teacher model and the actual data. Ensure that the distilled model maintains comparable accuracy and performance.

- Fine-tune the student model. After initial training, fine-tune the student model on the actual data to further improve its performance and adapt it to the specific characteristics of the target environment.

- Employ resource management. Optimize the deployment of the distilled model to make the best use of available computational resources, ensuring efficient operation in resource-constrained environments.

By following these best practices, organizations can effectively implement model distillation to create smaller, efficient models that maintain the performance of larger, complex models. This approach enables the deployment of high-performing machine learning models in environments with limited computational resources, making it valuable for various applications.

Techniques for Incremental Learning and Continuous Model Improvement

Incremental learning and continuous model improvement are essential for maintaining the performance and relevance of generative AI models over time. These techniques enable models to adapt to new data and evolving patterns without the need for complete retraining. This section delves into various techniques for achieving incremental learning and continuous model improvement. Incremental learning involves updating the model incrementally as new data becomes available rather than retraining it from scratch. This approach is particularly useful for applications with streaming data or where data is continuously generated.

Incremental learning algorithms update the model parameters based on the new data, allowing the model to adapt in real time to changes in the data distribution. This technique can significantly reduce the computational resources and time required for model updating.

The following are other important incremental learning and continuous model improvement.

- One of the key techniques for incremental learning is stochastic gradient descent (SGD). In SGD, the model parameters are updated incrementally based on small batches of data rather than the entire dataset. This approach allows the model to learn continuously from new data, making it well-suited for online learning scenarios. SGD is also computationally efficient, requiring less memory and processing power than batch gradient descent.

- Reinforcement learning is another technique that can facilitate continuous model improvement. In reinforcement learning, the model learns by interacting with its environment and receiving feedback in the form of rewards or penalties. This approach allows the model to continuously adapt its behavior based on the feedback it receives, leading to continuous improvement over time. Reinforcement learning is particularly useful for applications where the model needs to make sequential decisions, such as in robotics or autonomous systems.

- Meta-learning, or "learning to learn," is a technique that aims to improve the model's ability to learn from new data. In meta-learning, the model is trained on a variety of tasks, enabling it to quickly adapt to new tasks with minimal additional training. This approach can enhance the model's generalization capabilities and improve its performance on new, unseen data. Meta-learning is particularly useful for applications where the model must adapt to various tasks or environments.

- Curriculum learning is another technique that can aid in continuous model improvement. In curriculum learning, the model is trained on a sequence of tasks that gradually increase in complexity. This approach mimics the way humans learn, starting with simpler tasks and progressively tackling more challenging ones. Curriculum learning can help the model build a strong foundation and improve its performance on complex tasks over time.

- Self-supervised learning is a technique that leverages unlabeled data to improve the model's performance. In self-supervised learning, the model generates its own labels based on the structure of the data, allowing it to learn from vast amounts of unlabeled data. This approach can significantly enhance the model's ability to generalize to new data and improve its performance without the need for large amounts of labeled data.

- Lifelong learning, or continuous learning, is a technique that enables the model to retain knowledge from previous tasks while learning new ones. This approach helps to prevent the model from forgetting previously learned information, a phenomenon known as "catastrophic forgetting". Lifelong learning algorithms incorporate mechanisms to retain and recall knowledge from past tasks, allowing the model to build on its existing knowledge base and continuously improve over time.

- Adaptive learning rates are another technique that can facilitate continuous model improvement. Adaptive learning rate algorithms adjust the learning rate based on the model's performance, allowing faster convergence and improved accuracy. Techniques such as AdaGrad, RMSprop, and Adam are commonly used to implement adaptive learning rates, helping the model to learn more efficiently from new data.

- Transfer learning can also play a role in incremental learning and continuous model improvement. Transfer learning allows the model to quickly adapt to new tasks with minimal additional training by leveraging pre-trained models and fine-tuning them on new data. This approach can significantly reduce the time and resources required for model updating while improving the model's performance on new data.

In conclusion, efficient model retraining and updating strategies, along with incremental learning and continuous model improvement techniques, are essential for maintaining the performance and relevance of generative AI models. By implementing these strategies and techniques, organizations can ensure that their AI systems remain accurate, efficient, and adaptive to changing data and evolving patterns.

Responsible AI Model Retirement and Disposal

In the rapidly evolving landscape of artificial intelligence, the lifecycle of AI models extends beyond their development and deployment. As models become obsolete, it is crucial to implement responsible practices for their retirement and disposal. This chapter explores the importance of responsible AI model retirement and disposal, discussing best practices and techniques to minimize the environmental impact of retired models. From an environmental perspective, minimizing the carbon footprint is vital. This includes leveraging energy-efficient hardware accelerators, renewable energy sources, optimized model architectures, and techniques like pruning and multitenancy. Socially, rigorous debiasing of datasets, inclusive practices, content filtering, watermarking, and human oversight help mitigate risks around bias, privacy violations, and misinformation spread. Strong governance frameworks grounded in ethics and human rights must guide responsible GenAI development and deployment, embedding transparency, privacy protection, and human control mechanisms. The core tenets for building a strong, responsible AI practice for your generative AI applications are shown in Figure 9-1.

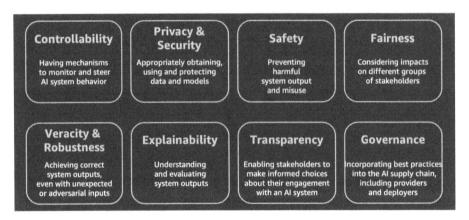

Figure 9-1. *Responsible AI tenets*

The responsible AI model retirement and disposal concept is rooted in the broader context of sustainable AI development. As organizations increasingly rely on AI technologies, the number of models being created, deployed, and eventually retired continues to grow. Without proper management, this cycle can lead to significant environmental, economic, and ethical challenges.

One of the primary considerations in responsible AI model retirement is determining when a model should be retired. This decision is often driven by factors such as declining performance, shifts in data distribution, or the emergence of more advanced models. Organizations must establish clear criteria and processes for evaluating model performance and relevance over time. Regular audits and performance assessments can help identify models that are no longer meeting the required standards or have become obsolete due to changes in the underlying data or business requirements.

Once the decision to retire a model has been made, organizations must consider the implications of removing it from active use. This process involves more than simply turning off the model; it requires careful planning and execution to ensure a smooth transition and minimize disruptions to dependent systems or processes. A phased approach to model retirement can be beneficial, gradually reducing the model's usage while simultaneously introducing its replacement or alternative solutions.

Data management is a critical aspect of responsible AI model retirement. Models often rely on vast amounts of data for training and operation. When retiring a model, organizations must carefully consider how to handle this associated data. This includes deciding whether to archive the data for future reference, repurpose it for other models, or securely delete it if it is no longer needed or if retention poses privacy risks. Proper data management during model retirement ensures compliance with data protection regulations and helps minimize the environmental impact by reducing unnecessary data storage.

Another important consideration in responsible AI model retirement is the preservation of knowledge. While a specific model may no longer be suitable for active use, the insights and knowledge gained during its development and operation can be valuable for future projects. Organizations should implement processes to document and preserve key learnings, including model architecture, training methodologies, and performance characteristics. This knowledge preservation supports future AI development efforts and contributes to the advancement of AI research and practices.

Transparency and communication are essential components of responsible AI model retirement. Stakeholders, including users, customers, and regulatory bodies, should be informed about the retirement of a model, especially if it has been used in critical decision-making processes. Clear communication about the reasons for retirement, the timeline for the transition, and any potential impacts can help maintain trust and ensure a smooth transition to alternative solutions.

As organizations retire AI models, they must also consider the ethical implications of this process. This includes assessing potential negative impacts on individuals or communities that may have relied on the model's outputs. For instance, if a model used in lending decisions is being retired, organizations must ensure that the transition does not inadvertently introduce biases or unfairly impact certain groups. Ethical considerations should guide the entire retirement process, from the initial decision to retire a model to implementing replacement solutions.

When it comes to the physical disposal of hardware associated with retired AI models, responsible practices are crucial to minimize environmental impact. Many AI models, particularly those used in large-scale applications, rely on specialized hardware such as high-performance GPUs or custom AI accelerators. These components often contain rare earth elements and other materials that can harm the environment if not properly disposed of.

Organizations should prioritize recycling and refurbishing hardware whenever possible. Many components used in AI systems can be repurposed for other computing tasks or recycled to recover valuable materials. Partnering with certified e-waste recycling facilities can ensure that hardware is disposed of in an environmentally responsible manner, minimizing the release of harmful substances into the environment. For organizations operating in cloud environments, responsible disposal may involve working with cloud service providers to ensure that resources allocated to retired models are properly decommissioned and repurposed. This can include releasing virtual machines, storage resources, and network allocations associated with the retired model.

To minimize the environmental impact of retired models, organizations should also consider the energy efficiency of their AI infrastructure.

- Energy-efficient hardware: Implementing energy-efficient hardware and optimizing data center operations can significantly reduce the carbon footprint associated with AI model lifecycles. This includes using renewable energy sources, implementing efficient cooling systems, and optimizing workload scheduling to maximize resource utilization.

- Model compression and optimization: Another technique for minimizing the environmental impact of retired models is to extend their useful life through model compression and optimization. Before retiring a model, organizations can explore techniques such as pruning, quantization, and knowledge distillation to create more efficient versions of the model. These compressed models may be suitable for deployment in resource-constrained environments or as part of edge computing solutions, effectively extending the model's lifecycle and reducing the need for frequent replacements.

- Model reuse: Organizations should also consider the potential for model reuse and transfer learning when retiring AI models. While a model may no longer be suitable for its original purpose, its learned features and knowledge may be valuable for other applications. Transfer learning techniques allow organizations to leverage pre-trained models as starting points for new tasks, reducing the computational resources required for training and potentially extending the useful life of the model's components.

- Lifecycle management strategy: Implementing a comprehensive lifecycle management strategy for AI models can significantly contribute to responsible retirement and disposal practices. This strategy should encompass all stages of a model's life, from development and deployment to maintenance and eventual retirement. By considering the entire lifecycle, organizations can make informed decisions about resource allocation, model architecture, and deployment strategies that facilitate more efficient and environmentally friendly retirement processes.

- Collaboration and knowledge sharing: Within the AI community, collaboration and knowledge sharing plays a crucial role in promoting responsible model retirement and disposal practices. Organizations can collectively work toward more sustainable AI development and deployment by sharing best practices, challenges, and solutions. Industry associations, academic institutions, and regulatory

bodies can contribute to developing standards and
guidelines for responsible AI model retirement,
ensuring that best practices are widely adopted across
the industry.

Conclusion

As AI technologies continue to advance, the importance of responsible
model retirement and disposal will only grow. Organizations must
stay informed about developments such as federated learning and
decentralized AI systems and be prepared to adapt their retirement
and disposal practices accordingly. In conclusion, responsible AI
model retirement and disposal are critical components of sustainable
AI development. By implementing thoughtful practices for model
evaluation, data management, knowledge preservation, and hardware
disposal, organizations can minimize the environmental impact of their AI
initiatives while maintaining ethical standards and regulatory compliance.
As the AI industry continues to evolve, a commitment to responsible
retirement and disposal practices will ensure the long-term sustainability
and societal benefit of AI technologies.

CHAPTER 10

Case Studies and Best Practices

Real-World Examples of Sustainable AI Implementations

In recent years, integrating artificial intelligence (AI) into various industries has led to significant advancements in efficiency, productivity, and sustainability. As organizations worldwide grapple with the urgent need to address environmental challenges, many have turned to AI as a powerful tool for implementing sustainable practices. This chapter explores real-world examples of sustainable AI implementations across different sectors, highlighting the challenges faced and the innovative solutions developed to overcome them.

The Energy Sector

The energy sector, a critical area for sustainability efforts, has seen remarkable progress through the application of AI. See Figure 10-1.

© Ishneet Kaur Dua and Parth Girish Patel 2024
I. K. Dua and P. G. Patel, *Optimizing Generative AI Workloads for Sustainability*,
https://doi.org/10.1007/979-8-8688-0917-0_10

Figure 10-1. *Representation of the energy sector*

A prime example is Google's collaboration with DeepMind to optimize cooling in their data centers. Data centers are notorious for their high energy consumption, particularly in cooling systems. The complex, dynamic environments of these facilities, with multiple variables affecting cooling efficiency, presented a significant challenge. Google and DeepMind rose to this challenge by implementing a deep learning system that analyzes data from thousands of sensors throughout the data centers. This AI system developed predictive models to optimize cooling parameters in real time and created a control system that autonomously adjusts cooling settings. The results of this implementation were nothing short of impressive. Google achieved a 40% reduction in energy used for cooling and a 15% decrease in overall power usage effectiveness. These improvements led to significant cost savings and substantially reduced the company's carbon footprint. The success of this project demonstrates the potential for AI to dramatically improve energy efficiency in data centers and other energy-intensive industries, contributing to global efforts to combat climate change.

The Agricultural Sector

In the agricultural sector, sustainable AI implementations are revolutionizing farming practices. See Figure 10-2.

Figure 10-2. *Representation of the agricultural sector*

Blue River Technology, now a part of John Deere, developed an AI-powered precision spraying system called See & Spray. This innovation addresses several critical challenges in modern agriculture, including the overuse of herbicides leading to environmental pollution, the development of herbicide resistance in weeds, and the inefficient practice of blanket spraying entire fields. The See & Spray system employs computer vision and machine learning to precisely identify and target individual weeds. By developing real-time decision-making algorithms for precise herbicide application, Blue River Technology created a system that can accurately distinguish between crops and weeds. This targeted approach has resulted in up to 90% reduction in herbicide use, improved crop yields due to more effective weed control, and significant cost savings for farmers. The impact of this technology on sustainability in agriculture is profound. By dramatically reducing chemical runoff into water systems, decreasing the risk of herbicide resistance development, and promoting more environmentally friendly farming practices, the See & Spray system represents a significant step toward sustainable agriculture. As this technology continues to evolve and become more widely adopted, it has the potential to transform farming practices on a global scale, reducing the environmental impact of agriculture while improving food security.

The Transportation Sector

The transportation sector, a major contributor to global carbon emissions, is another area where sustainable AI implementations are making a significant impact. See Figure 10-3.

Figure 10-3. *Representation of the transportation sector*

The transportation sector, a major contributor to global carbon emissions, is another area where sustainable AI implementations are making a significant impact. Waymo, a subsidiary of Alphabet Inc., has been at the forefront of developing autonomous driving technology to make transportation safer and more efficient. The challenges in this field include reducing carbon emissions from traditional vehicles, alleviating traffic congestion, improving route planning efficiency, and addressing safety concerns inherent in autonomous driving. Waymo's approach involves implementing advanced AI algorithms for real-time decision-making and route optimization. They have developed sophisticated sensor fusion techniques to accurately perceive the environment and created machine learning models capable of predicting and responding to various traffic scenarios. While the technology is still in development, early results are promising. Autonomous vehicles have the potential to reduce fuel consumption by 15%–20% through optimized driving, improve traffic flow, reduce congestion, and enhance safety through AI-powered perception and decision-making.

The sustainability impact of widespread autonomous vehicle adoption could be substantial. By reducing carbon emissions through efficient driving and route optimization, increasing the potential for adoption of electric vehicles in autonomous fleets, and improving urban mobility while reducing the need for extensive parking infrastructure, autonomous driving technology represents a significant step toward sustainable transportation systems.

The Manufacturing Sector

In the manufacturing sector, Siemens has demonstrated the power of AI in optimizing industrial processes for sustainability. See Figure 10-4.

Figure 10-4. *Representation of the manufacturing sector*

The company implemented an AI system to enhance the performance and efficiency of gas turbines used in power generation. This application addressed several challenges, including inefficient fuel consumption in gas turbines, balancing power output with emissions reduction, and the complex operational parameters affecting turbine performance. Siemens' solution involved developing neural network models to predict and optimize turbine performance, implementing reinforcement learning algorithms for real-time control adjustments, and creating a digital twin system for simulating and testing optimization strategies. The results were impressive, with a 10%–15% reduction in nitrogen oxide emissions, a 1%–2% improvement in fuel efficiency, and extended maintenance intervals that reduced downtime.

The sustainability impact of this AI implementation extends beyond the immediate improvements in emissions and efficiency. By demonstrating the potential for AI to optimize complex industrial processes, Siemens has paved the way for similar applications across various industries, potentially leading to significant reductions in industrial emissions and resource consumption on a global scale.

The Retail Sector

While not typically associated with heavy industry or energy consumption, the retail sector has also found innovative ways to leverage AI for sustainability. See Figure 10-5.

Figure 10-5. *Representation of the retail sector*

Walmart's development and implementation of Eden, an AI-powered system for managing fresh produce quality and reducing food waste, is an excellent example. The challenges in this area included high levels of food waste in the retail supply chain, inconsistent quality of fresh produce, and the complex logistics involved in managing perishable goods. Walmart's solution involved implementing machine learning algorithms to predict freshness and shelf life, developing computer vision systems to assess produce quality, and creating an AI-driven supply chain management system to optimize inventory and distribution. The results were remarkable, with $86 million saved in the first year through reduced food waste, improved produce quality for customers, and more efficient inventory management that reduced overstocking.

The sustainability impact of this implementation is significant. By substantially reducing food waste, Walmart has decreased the carbon emissions associated with producing, transporting, and disposing spoiled produce. Moreover, this more efficient use of agricultural resources contributes to more sustainable food systems. As similar technologies are adopted across the retail sector, the cumulative impact on reducing food waste and its associated environmental costs could be substantial.

The Healthcare Sector

In the healthcare sector, AI is being leveraged to improve the efficiency of medical processes and patient outcomes, with positive implications for sustainability. See Figure 10-6.

Figure 10-6. *Healthcare sector*

Zebra Medical Vision's development of AI algorithms to assist radiologists in analyzing medical images is a prime example. The challenges in this field included the high workload for radiologists, leading to potential burnout and errors, increasing demand for medical imaging services, and the need for early detection of diseases to improve patient outcomes. Zebra Medical Vision's solution involved implementing deep learning algorithms for analyzing various types of medical images, developing AI models to detect and prioritize urgent cases, and creating a system for automated reporting and workflow optimization. The results have been impressive, with 91% accuracy in detecting breast cancer in mammograms, a 95% reduction in time to diagnose brain bleeds, and improved early detection rates for various diseases.

The sustainability impact of this AI implementation in healthcare is multifaceted. By enabling more efficient use of healthcare resources, reducing the need for repeat scans (thereby lowering radiation exposure and energy consumption), and improving patient outcomes through early detection and treatment, this technology contributes to a more sustainable healthcare system. As these AI systems become more widespread, they have the potential to significantly reduce the environmental footprint of healthcare while improving public health outcomes.

The Financial Sector

While not typically associated with direct environmental impact, the financial sector has also found ways to leverage AI for sustainability. See Figure 10-7.

Figure 10-7. *Financial sector*

JP Morgan's development of the Contract Intelligence (COiN) platform demonstrates how AI can improve efficiency and reduce resource consumption in unexpected ways. The challenges in this area included time-consuming manual review of financial documents, high risk of human error in document analysis, and inefficient use of skilled legal professionals' time. JP Morgan's solution involved implementing natural language processing to extract key information from documents, developing machine learning models to classify and analyze contract terms, and creating an AI-driven system for identifying anomalies and potential risks in agreements. The results were significant, with 360,000 hours of manual work saved annually, the ability to review 12,000 commercial credit agreements in seconds, and improved accuracy and consistency in document analysis.

While the sustainability impact might seem less direct in this case, it is nonetheless significant. By reducing paper usage through more efficient document processing, decreasing energy consumption associated with manual document review, and enabling more efficient use of human resources (potentially reducing office space requirements), the COiN platform contributes to sustainability in the financial sector. As similar technologies are adopted more widely, the cumulative effect on resource consumption in office-based industries could be substantial.

The Water Management

In the critical area of water management, AI is being leveraged to address one of the most pressing sustainability challenges of our time. See Figure 10-8.

Figure 10-8. *Water management for sustainability*

Aquasight's development of an AI-powered platform to help water utilities optimize their operations and reduce water waste is a prime example. The challenges in this field included high levels of water loss in distribution systems, inefficient energy use in water treatment and distribution, and difficulty in predicting and managing water demand. Aquasight's solution involved implementing machine learning algorithms to detect leaks and anomalies in water networks, developing predictive models for water demand and quality management, and creating an AI-driven system for optimizing pump operations and energy use. The results have been impressive, with up to 25% reduction in non-revenue water loss, a 20% decrease in energy consumption for water distribution, and improved water quality management and regulatory compliance.

The sustainability impact of this AI implementation in water management is profound. By conserving water resources, reducing energy consumption (and associated carbon emissions) in water distribution, and enabling more efficient use of chemical treatments in water processing, this technology directly addresses some of the most critical sustainability challenges related to water resources. As water scarcity becomes an increasingly pressing issue globally, the widespread adoption of such AI-driven optimization systems could play a crucial role in ensuring sustainable water management.

The Waste Management Sector

The waste management sector, another critical area for sustainability efforts, has also seen innovative applications of AI. See Figure 10-9.

Figure 10-9. *Waste management for sustainability*

AMP Robotics' development of AI-powered robots to improve the efficiency and accuracy of recycling sorting processes addresses several key challenges in this field. These include the inefficient and often inaccurate manual sorting of recyclables, contamination of recycling streams reducing the value of recycled materials, and high labor costs coupled with challenging working conditions in recycling facilities. AMP Robotics' solution involved implementing computer vision and machine learning for real-time identification of recyclable materials, developing robotic systems for high-speed, accurate sorting, and creating an AI model that can learn to identify new types of materials over time. The results have been remarkable, with the robots capable of up to 70 picks per minute (twice the speed of human sorters), 99% accuracy in material identification, and the ability to operate 24/7, significantly increasing overall recycling capacity.

The sustainability impact of this AI implementation in waste management is significant. By increasing recycling rates, improving the quality of recycled materials, reducing contamination in recycling streams, and enabling more efficient use of resources through improved recycling processes, this technology directly contributes to the circular economy. As these AI-powered recycling systems become more widespread, they have the potential to dramatically improve the efficiency and effectiveness of recycling efforts globally, reducing the environmental impact of waste and conserving valuable resources.

The Forestry Sector

In the forestry sector, AI is being leveraged to address one of the most critical aspects of climate change mitigation: carbon sequestration. See Figure 10-10.

Figure 10-10. *Forestry sector*

Pachama's development of an AI system to monitor and verify carbon capture in forests improves the accuracy and efficiency of carbon offset programs. The challenges in this area included the difficulty in accurately measuring carbon sequestration in forests, time-consuming and expensive manual forest surveys, and the need for ongoing monitoring to ensure the permanence of carbon offsets. Pachama's solution involved implementing machine learning algorithms to analyze satellite imagery and LiDAR (light detection and ranging) data, developing AI models to estimate forest biomass and carbon content, and creating a system for continuous monitoring and verification of forest carbon stocks. The results have been impressive, with up to 90% cost reduction compared to traditional forest carbon measurement methods, improved accuracy in carbon stock estimates, and the ability to monitor large forest areas in near real-time.

The sustainability impact of this AI implementation in forestry is far-reaching. By enhancing the credibility and effectiveness of forest carbon offset programs, improving forest conservation efforts through better monitoring, and enabling more efficient allocation of resources for climate change mitigation, this technology plays a crucial role in global efforts to combat climate change. As the importance of forest conservation and reforestation in climate change mitigation becomes increasingly recognized, the widespread adoption of such AI-driven monitoring and verification systems could significantly enhance the effectiveness of these efforts.

In conclusion, these real-world examples demonstrate the diverse and impactful applications of sustainable AI across various industries. From optimizing energy use in data centers to revolutionizing agriculture, transforming transportation to enhancing recycling processes, AI is a powerful tool in addressing some of the most pressing sustainability challenges of our time. While challenges remain in implementing and scaling these solutions, the potential for AI to contribute to sustainability goals is significant and growing.

The examples discussed in this chapter represent just the beginning of what is possible with sustainable AI implementations. As more organizations recognize the potential of AI in addressing sustainability challenges, and as the technology continues to evolve, we can anticipate a future where AI plays an increasingly central role in creating a more sustainable world. The journey toward this future requires ongoing innovation, collaboration across sectors, and a steadfast commitment to leveraging technology for the betterment of our planet and society as a whole.

Lessons Learned and Best Practices from Sustainable AI Implementations

The case studies and real-world implementations of sustainable AI across various industries provide valuable insights into the challenges, solutions, and best practices for leveraging artificial intelligence to address environmental and social issues. This chapter summarizes the key lessons learned and highlights best practices for sustainable AI implementation and optimization.

- Cross-industry applicability: One of the most striking lessons from these case studies is the broad applicability of AI in addressing sustainability challenges across diverse industries. From energy and agriculture to transportation and healthcare, AI has demonstrated its potential to drive significant improvements in efficiency, resource utilization, and environmental impact. This cross-industry applicability underscores the versatility of AI as a tool for sustainability and highlights the importance of knowledge sharing and collaboration across sectors.

- Data-driven decision-making: A common thread among successful sustainable AI implementations is the emphasis on data-driven decision-making. In cases such as Google's data center cooling optimization and Walmart's fresh produce management, the ability to collect, analyze, and act upon vast amounts of data in real time has been crucial to achieving sustainability goals. This lesson emphasizes the importance of robust data collection and management systems as a foundation for effective AI implementation.

- Balancing efficiency and sustainability: Many case studies demonstrate that improving efficiency and promoting sustainability are often complementary goals. For instance, Siemens' AI-powered gas turbine optimization reduced emissions and improved fuel efficiency and extended maintenance intervals. This lesson highlights the potential for sustainable AI implementations to deliver environmental and economic benefits, creating a strong business case for adoption.

- Continuous learning and adaptation: The dynamic nature of environmental challenges and rapid technological advancement necessitate AI systems that can learn and adapt continuously. Examples such as AMP Robotics' recycling robots, which can learn to identify new types of materials over time, illustrate the importance of designing AI systems with the capacity for ongoing learning and improvement.

- Human–AI collaboration: While AI has shown remarkable capabilities in many areas, the case studies also highlight the continued importance of human expertise and oversight. In healthcare, for instance, Zebra Medical Vision's AI assists radiologists rather than replacing them, demonstrating the value of human–AI collaboration in achieving optimal results.

- Scalability and replicability: Many successful implementations, such as Blue River Technology's See & Spray system in agriculture, demonstrate the potential for scalability and replicability across different contexts. This lesson underscores the importance of designing AI solutions with scalability to maximize their impact on sustainability.

- A holistic approach to sustainability: The case studies reveal that the most effective sustainable AI implementations take a holistic approach, considering direct environmental impacts, indirect effects, and long-term consequences. For example, Pachama's forest carbon verification system addresses not just immediate carbon sequestration but also long-term forest conservation and the credibility of carbon offset programs.

Best Practices for Sustainable AI Implementation and Optimization

- Sustainability goals: Establish clear, measurable sustainability goals at the outset of any AI implementation project. These goals should align with broader organizational and societal sustainability objectives and provide a clear framework for measuring success.

- Comprehensive data strategy: Develop a comprehensive data strategy that addresses data collection, quality, storage, and accessibility. Ensure that data practices are sustainable, considering energy consumption in data centers and the environmental impact of data collection methods.

- Ethical considerations: Integrate ethical considerations into every stage of AI development and implementation. This includes addressing issues of bias, fairness, and transparency and considering the broader societal impacts of AI systems.

- Interdisciplinary collaboration: Foster collaboration between AI experts, domain specialists, and sustainability professionals. This interdisciplinary approach ensures that AI solutions are technically sound and aligned with sustainability principles.

- Lifecycle assessment: Conduct thorough lifecycle assessments of AI systems, considering their environmental impact from development through deployment and eventual decommissioning. This practice helps ensure that the overall impact of AI implementation is net positive for sustainability.

- Adaptive and flexible design: Design AI systems with adaptability and flexibility in mind, allowing for continuous improvement and adjustment to changing environmental conditions and sustainability requirements.

- Stakeholder engagement: Engage with a wide range of stakeholders, including employees, customers, local communities, and regulatory bodies, throughout the AI implementation process. This engagement can provide valuable insights, build trust, and ensure that AI solutions address the needs and concerns of all affected parties.

- Transparency and explainability: Prioritize transparency and explainability in AI systems, particularly when making decisions with significant environmental or social impacts. This practice builds trust and allows for better oversight and improvement of AI systems.

- Continuous monitoring and evaluation: Implement robust systems for continuous monitoring and evaluation of AI performance against sustainability metrics. Regular assessments allow for timely adjustments and ensure that AI systems continue to meet sustainability goals over time.

- Knowledge sharing and open innovation: Participate in knowledge-sharing initiatives and open innovation platforms to accelerate developing and adopting sustainable AI solutions across industries.

- Energy-efficient AI: Prioritize energy efficiency in AI system design and deployment, considering hardware and software optimizations. This includes exploring edge computing solutions where appropriate to reduce data transmission and centralized processing requirements.

- Integration with existing systems: Design AI solutions that integrate seamlessly with existing systems and processes to minimize disruption and maximize adoption. This may involve phased implementation approaches and comprehensive training programs for users.

- Resilience and redundancy: Build resilience and redundancy into AI systems, particularly those critical to environmental monitoring or resource management. This ensures continued operation and effectiveness even in the face of unexpected challenges or system failures.

- Regulatory compliance and anticipation: Stay abreast of and comply with relevant regulations related to both AI and sustainability. Additionally, anticipate future regulatory changes and design AI systems with the flexibility to adapt to evolving legal and policy landscapes.

- Measurable impact metrics: Develop and track clear, measurable impact metrics directly related to sustainability goals. These metrics should go beyond traditional business KPIs to include specific environmental and social impact indicators.

In conclusion, the lessons learned from real-world sustainable AI implementations highlight AI's tremendous potential in addressing global sustainability challenges and the complexities involved in realizing this potential. By adhering to best practices that emphasize clear goals, comprehensive data strategies, ethical considerations, and continuous improvement, organizations can harness the power of AI to drive meaningful progress toward a more sustainable future.

The journey toward sustainable AI implementation is ongoing, and these lessons and best practices will undoubtedly evolve as we gain more experience and technology continues to advance. However, by building on the successes and learning from the challenges faced in these pioneering efforts, we can accelerate the development and adoption of AI solutions that drive business value and contribute significantly to global sustainability goals.

As we move forward, it will be crucial to balance innovation and responsibility, ensuring that our pursuit of technological advancement aligns with our commitment to environmental stewardship and social well-being. By doing so, we can harness the transformative power of AI to create a more sustainable, equitable, and prosperous world for current and future generations.

Future Trends and Research Directions in Sustainable AI Practices

As the field of artificial intelligence continues to evolve rapidly, so does its intersection with sustainability. The coming years promise exciting developments in sustainable AI practices, with numerous areas ripe for exploration and innovation. This chapter examines the trends and research directions that will likely shape the landscape of sustainable AI.

One of the most promising trends in sustainable AI is the development of more energy-efficient AI models. As AI systems become increasingly complex and widespread, their energy consumption has become a significant concern. Future research will likely focus on creating AI architectures that deliver high performance while minimizing energy use. This may involve innovations in hardware design, such as neuromorphic computing systems that mimic the energy efficiency of biological brains, or advancements in software optimization techniques that can reduce the computational requirements of AI models.

Another important trend is the integration of AI with renewable energy systems. See Figure 10-11.

Figure 10-11. *Solar panels for clean energy sources*

As the world transitions toward cleaner energy sources, AI will play a crucial role in optimizing renewable energy generation, distribution, and consumption. Future research may explore how AI can improve the forecasting of renewable energy production, enhance grid stability with high penetration of intermittent sources, and optimize energy storage systems. These advancements could significantly accelerate the adoption of renewable energy and contribute to global efforts to combat climate change.

The application of AI in circular economy initiatives is another area poised for growth. See Figure 10-12.

Figure 10-12. *Circular economy for sustainability*

As societies grapple with resource scarcity and waste management challenges, AI can help design products for easier recycling, optimize waste sorting processes, and identify new material reuse opportunities. Future research may focus on developing AI systems that can track and trace materials throughout their lifecycle, predict optimal product refurbishment or recycling times, and even suggest innovative ways to repurpose waste materials.

In biodiversity conservation, AI is set to play an increasingly important role. See Figure 10-13.

Figure 10-13. *Biodiversity conservation*

Future trends may include developing more sophisticated AI systems for monitoring ecosystems, tracking wildlife populations, and predicting the impacts of climate change on biodiversity. Research may also explore how AI can assist in the design and management of protected areas, help combat illegal wildlife trade, and support the restoration of degraded ecosystems. The intersection of AI and sustainable agriculture is another area ripe for innovation. While AI is already being used in precision farming, future research may focus on developing more advanced systems for crop disease detection, optimizing water and nutrient use, and predicting crop yields under various climate scenarios. There may also be increased emphasis on using AI to support regenerative agriculture practices and enhance food systems' resilience in the face of climate change.

In the urban environment, the concept of "smart cities" is likely to evolve with advancements in sustainable AI. See Figure 10-14.

Figure 10-14. *Smart cities for sustainability*

Future research may explore how AI can optimize urban resource use, improve air quality monitoring and management, enhance urban biodiversity, and create more efficient and sustainable transportation systems. There may also be increased focus on using AI to model and mitigate the urban heat island effect, a growing concern as cities face rising temperatures due to climate change.

Sustainable finance is another area where AI is poised to make significant contributions. Future trends may include developing more sophisticated AI systems for assessing the environmental and social impacts of investments, predicting climate-related financial risks, and optimizing sustainable investment portfolios. Research may also explore how AI can enhance the transparency and accountability of corporate sustainability reporting, potentially revolutionizing how we measure and value sustainable business practices.

As AI systems become more prevalent in sustainability applications, there will likely be increased focus on ensuring the fairness, transparency, and accountability of these systems. Future research may explore methods for detecting and mitigating bias in AI systems used for environmental decision-making, developing explainable AI models for sustainability applications, and creating governance frameworks for the ethical use of AI in environmental management.

Integrating AI with other emerging technologies, such as blockchain and the Internet of Things (IoT), presents exciting possibilities for sustainable practices. See Figure 10-15.

Figure 10-15. *Blockchain and the IoT for sustainability*

Future research may explore how these technologies can work together to create more transparent and efficient supply chains, enable peer-to-peer renewable energy trading, or support community-based environmental monitoring initiatives.

Another important area for future research is the development of AI systems that can help societies adapt to the impacts of climate change. This may include advanced climate modeling and prediction systems, AI-powered early warning systems for extreme weather events, and decision support tools for climate adaptation planning. There may also be increased focus on using AI to optimize the design and operation of climate-resilient infrastructure.

The use of AI in environmental education and behavior change is another promising area for exploration. Future research may investigate how AI can be used to create personalized environmental education programs, gamify sustainable behaviors, or provide real-time feedback on individual and collective environmental impacts. These applications could play a crucial role in fostering a culture of sustainability and empowering individuals to make more environmentally conscious decisions.

As the field of quantum computing advances, there may be opportunities to leverage this technology for sustainable AI applications. See Figure 10-16.

Figure 10-16. *Quantum computing*

Future research may explore how quantum algorithms can solve complex optimization problems in areas such as energy systems, materials science, and climate modeling, potentially leading to breakthroughs in sustainable technology development.

Finally, there is likely to be increased focus on developing AI systems to assist with global environmental governance and policy-making. This may include AI tools for analyzing and synthesizing vast amounts of environmental data, modeling the potential impacts of different policy scenarios, and facilitating international cooperation on environmental issues.

Conclusion

In conclusion, the future of sustainable AI practices is rich with possibilities. From energy-efficient AI architectures to advanced climate modeling systems, from circular economy applications to biodiversity conservation tools, the potential for AI to contribute to global sustainability efforts is immense. However, realizing this potential requires continued research, innovation, and collaboration across disciplines and sectors.

As we move forward, it will be crucial to ensure that ethical considerations and a commitment to social equity guide the development of sustainable AI practices. The challenge lies not just in creating more advanced AI systems but in developing solutions that are accessible, beneficial, and sustainable for all of humanity and the planet we inhabit.

The journey toward truly sustainable AI is just beginning, and the coming years promise to be an exciting time of discovery and innovation. By continuing to push the boundaries of what's possible while remaining grounded in the principles of sustainability and ethical responsibility, we can harness the power of AI to create a more sustainable and equitable world for current and future generations.

Index

A

Adaptive Replacement
Cache (ARC), 158
AI, *see* Artificial intelligence (AI)
Amazon Web Services (AWS),
28, 75
distributed/parallel training
strategies, 180
hybrid architectures, 222, 223
hyperparameters/training
schedules, 188
inference optimization, 196
renewable energy, 128
AMP Robotics' development, 301
Application performance
management (APM), 165
Application-specific integrated
circuit (ASIC), 64
Aquasight's development, 299
ARC, *see* Adaptive Replacement
Cache (ARC)
Artificial intelligence (AI)
containerization (*see*
Containerization
technologies)
environmental impacts
benefits, 31

CO_2 emissions footprint, 34
ecological impacts, 32
emissions/electronics
waste, 32
environmental costs, 35
generative models, 36
hazardous chemicals, 34
high-impact domains, 33
immense energy, 33
model parameter, 32
neural networks, 31
training phase, 34
hierarchy, 1, 2
lifecycle management (*see*
Lifecycle
management (AI))
sustainable AI (*see* Real-
world (AI)
implementations)
workload (*see* Workload
management, AI)
ASIC, *see* Application-specific
integrated circuit (ASIC)
Autoscaling and load balancing
techniques
AI-driven optimization, 254
challenges/considerations, 254
comprehensive monitoring, 253

H

T

GPSR Compliance
The European Union's (EU) General Product Safety Regulation (GPSR) is a set
of rules that requires consumer products to be safe and our obligations to
ensure this.

If you have any concerns about our products, you can contact us on

ProductSafety@springernature.com

In case Publisher is established outside the EU, the EU authorized
representative is:

Springer Nature Customer Service Center GmbH
Europaplatz 3
69115 Heidelberg, Germany

www.ingramcontent.com/pod-product-compliance
Lightning Source LLC
LaVergne TN
LVHW051637050326
832903LV00022B/786

* 9 7 9 8 8 6 8 8 0 9 1 6 3 *